PRAI

ENTANGLED MINDS

"In this triumph of scientific imagination, Dean Radin shows in clear language how the mysteries of psychology and the mysteries of quantum mechanics may combine to point to a 'new reality' that makes the most daring science fiction look tame by comparison."

—Michael Grosso, Ph.D., philosopher, author of *Experiencing the Next World Now*

"The implications of Radin's premises are majestic. His views strike at the heart of the notion of the isolated individual, and replace it with an image of the unity of all minds. Radin shows that togetherness does not have to be developed; it already exists and needs only to be realized. In a world seething with enmity, difference, and white-knuckled competition, it is difficult to imagine a more important contribution."

—Larry Dossey, M.D., author of *The Extraordinary Healing Power of Ordinary Things*

"As scientists we tend a precarious campfire in the midst of the great Jungle of Ignorance. Dean Radin's work is a reliable and trustworthy guide to important shapeshifting phenomena lying just outside that comfortable circle of light."

—Nick Herbert, Ph.D., physicist, author of *Quantum Reality*

. . . and for Dean Radin's Award-Winning Bestseller

THE CONSCIOUS UNIVERSE

"A brilliant book."

—Gertrude Schneider, Ph.D., psychologist,
Journal of the American Society for Physical Research

"[An] intriguing, exhaustive tome."

—*Entertainment Weekly* (rated B+)

"The extensive data and sober arguments . . . show that psi research is worthy of consideration."

—MSNBC's "The Site"

"I loved it. It made my head spin for days."

—Scott Adams, cartoonist and author of *The Dilbert Principle*

"Cutting perceptively through the spurious arguments frequently made by skeptics, Radin shows that the evidence in favor of [paranormal] existence is overwhelming."

—Brian Josephson, Ph.D., Nobel Laureate
and professor of physics, Cambridge University

ALSO BY DEAN RADIN:

The Conscious Universe: The Scientific Truth of Psychic Phenomena

ENTANGLED MINDS

Extrasensory Experiences in a Quantum Reality

DEAN RADIN

PARAVIEW POCKET BOOKS
New York   London   Toronto   Sydney

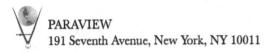 PARAVIEW
191 Seventh Avenue, New York, NY 10011

 POCKET BOOKS, a division of Simon & Schuster, Inc.
1230 Avenue of the Americas, New York, NY 10020

Library of Congress Cataloging-in-Publication Data

Radin, Dean I.
    Entangled minds: extrasensory experiences in a quantum reality / Dean
Radin.
        p. cm.
    Includes bibliographical references and index.
    1. Parapsychology. 2. Quantum theory—Miscellanea. I. Title

BF1040.R33 2006
133.8—dc22   2005058657

ISBN-13: 978-1-4165-1677-4
ISBN-10:   1-4165-1677-8

This Paraview Pocket Books trade paperback edition April 2006

10 9 8 7 6 5 4 3 2 1

Manufactured in the United States of America

For information regarding special discounts for bulk purchases,
please contact Simon & Schuster Special Sales at 1-800-456-6798
or business@simonandschuster.com.

TO SUSIE

CONTENTS

	Preface	1
CHAPTER 1:	In the Beginning	5
CHAPTER 2:	Naked Psi	21
CHAPTER 3:	Who Believes?	35
CHAPTER 4:	Origins	52
CHAPTER 5:	Putting Psi to the Test	81
CHAPTER 6:	Conscious Psi	98
CHAPTER 7:	Unconscious Psi	131
CHAPTER 8:	Gut Feelings	142
CHAPTER 9:	Mind-Matter Interaction	146
CHAPTER 10:	Presentiment	161
CHAPTER 11:	Gaia's Dreams	181
CHAPTER 12:	A New Reality	208
CHAPTER 13:	Theories of Psi	240
CHAPTER 14:	Next	275
	Acknowledgments	297
	Endnotes	301
	Index	343

PREFACE

If you do not get *schwindlig* [dizzy] sometimes when you think about
these things then you have not really understood it [quantum theory].

Niels Bohr

One of the most surprising discoveries of modern physics is that
objects aren't as separate as they may seem. When you drill
down into the core of even the most solid-looking material, sep-
arateness dissolves. All that remains, like the smile of the
Cheshire Cat from *Alice in Wonderland,* are relationships extending curiously throughout space and time. These connections
were predicted by quantum theory and were called "spooky action at a distance" by Albert Einstein. One of the founders of
quantum theory, Erwin Schrödinger, dubbed this peculiarity *entanglement,* and said "I would not call that *one* but rather *the* characteristic trait of quantum mechanics."

The deeper reality suggested by the existence of entanglement is so unlike the world of everyday experience that until recently, many physicists believed it was interesting only for
abstract theoretical reasons. They accepted that the microscopic
world of elementary particles could become curiously entangled, but those entangled states were assumed to be fleeting and

have no practical consequences for the world as we experience it. That view is rapidly changing.

Scientists are now finding that there are ways in which the effects of microscopic entanglements "scale up" into our macroscopic world. Entangled connections between carefully prepared atomic-sized objects can persist over many miles. There are theoretical descriptions showing how tasks can be accomplished by entangled groups without the members of the group communicating with each other in any conventional way. Some scientists suggest that the remarkable degree of coherence displayed in living systems might depend in some fundamental way on quantum effects like entanglement. Others suggest that conscious awareness is caused or related in some important way to entangled particles in the brain. Some even propose that the entire universe is a single, self-entangled object.

If these speculations are correct, then what would human experience be like in such an interconnected universe? Would we occasionally have numinous feelings of connectedness with loved ones, even at a distance? Would such experiences evoke a feeling of awe that there's more to reality than common sense implies? Could "entangled minds" be involved when you hear the telephone ring and somehow know—instantly—who's calling? If we did have such experiences, could they be due to real information that somehow bypassed the usual sensory channels? Or are such reports better understood as coincidences or delusions?

These are the types of questions explored in this book. We'll find that there's substantial experimental evidence for a few types of genuine psi phenomena. And we'll learn why, until very recently, science has largely ignored these interesting effects. For centuries, scientists assumed that everything can be explained by mechanisms analogous to clockworks. Then, to everyone's surprise, over the course of the twentieth century we learned that this commonsense assumption is wrong. When the fabric of reality is examined very closely, nothing resembling clockworks can be found. Instead, reality is woven from strange, "holistic" threads that aren't located precisely in space or time. Tug on a dangling loose end from this fabric of reality, and the whole cloth twitches, instantly, throughout all space and time.

Science is at the very earliest stages of understanding entanglement, and there is much yet to learn. But what we've seen so far provides a new way of thinking about psi. No longer are psi experiences regarded as rare human talents, divine gifts, or "powers" that magically transcend ordinary physical boundaries. Instead, psi becomes an unavoidable consequence of living in an interconnected, entangled physical reality. Psi is reframed from a bizarre anomaly that doesn't fit into the normal world—and hence is labeled *paranormal*—into a natural phenomenon of physics.

The idea of the universe as an interconnected whole is not new; for millennia it's been one of the core assumptions underlying Eastern philosophies. What is new is that Western science is slowly beginning to realize that *some* elements of that ancient lore might have been correct. Of course, adopting a new ontology is not to be taken lightly. When it comes to serious topics like one's view of reality, it's sensible to adopt the conservative maxim, "if it ain't broke, don't fix it." So we're obliged to carefully examine the evidence and see if psi is real or not. If the conclusion is positive, then previous assumptions about the relationship between mind and matter are wrong and we'll need to come up with alternatives.

As we explore the concept of psi as "entangled minds," we'll consider examples of psi experiences in life and lab, we'll take a survey of the origins of psi research, we'll explore the outcomes of thousands of controlled laboratory tests, and we'll debunk some skeptical myths. Then we'll explore the fabric of reality as revealed by modern physics and see why it's becoming increasingly relevant to understanding why and how psi exists. At the end, we'll find that the nineteenth century English poet Francis Thompson may have said it best:

All things by immortal power,
Near and Far
Hiddenly
To each other linked are,
That thou canst not stir a flower
Without troubling of a star

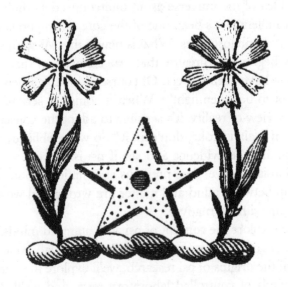

CHAPTER 1
iN THE BEGiNNiNG

Reality isn't what it used to be.

Historians disagree over precisely when it changed. Some say the year 1905. Others point to key events in 1964 and 1982. I think of it as a glacial shift that took most of the twentieth century.

The new reality is not yet fully understood. But what we've grasped so far suggests a startling metamorphosis of the old, something so unexpected that for many decades hardly anyone paid attention to it. Those who did notice were, at first, speechless. When they were able speak again, they muttered terms like *shocking, bizarre, mind-boggling,* and *spooky.* "They" are not advertisers hyping commercial products we don't need, but sedate physicists and philosophers struggling to understand what the new reality means.

"New reality" refers to the modern understanding of the interconnected medium in which we live, the fabric of reality as revealed by modern physics. The purpose of this book is to explore one of the consequences of this new reality for the familiar realm of human experience. We'll see what happens when experience is viewed from the perspective of quantum theory, and in doing so we'll find a surprise: Certain phenomena previously thought to be impossible might, in fact, exist.

I am speaking of psychic phenomena. Telepathy. Clairvoyance. Psychokinesis.

Some may object that linking the elegance of quantum theory to the spookiness of psychic phenomena is illegitimate, that it's a mistake to claim a connection exists simply because these two domains are permeated with uncanny effects. This objection is certainly understandable. Quantum theory is a mathematically precise and exquisitely well-tested description of the observable world. Psychic phenomena are slippery, subjective events with a checkered past. But as it turns out, the fabric of reality suggested by quantum theory and the observations associated with psychic phenomena bear striking resemblances. They are eerily weird in *precisely* the right way to suggest a meaningful relationship. That's the connection we'll explore here: The *psi* connection.

The term psi was coined as a neutral term for psychic phenomena in 1942 by British psychologist Robert Thouless. It refers to the 23rd letter of the Greek alphabet and is pronounced "sigh." Psi is also the first letter of the Greek word *psyche,* meaning soul or mind. Thouless chose this term as a way to refer to psychic experience without implying origins or mechanisms. Common psi experiences include mind-to-mind connections (telepathy), perceiving distant objects or events (clairvoyance), perceiving future events (precognition), and mind-matter interactions (psychokinesis). Psi may also be involved in intuitive hunches, gut feelings, distant healing, the power of intention, and the sense of being stared at.

There are words for psi experiences in every language, from Arabic to Zulu, Czech to Manx Gaelic.[1] The universality of the words reflects the fact that these phenomena are basic to human experience. And indeed psi experiences have been reported by people in all cultures, throughout history, and at all ages and educational levels.

FORBIDDEN SCIENCE

The general public has always been interested in psi phenomena. But within the scientific orthodoxy psi has been regarded as either a genuine hot potato or a Mr. Potato Head toy. Many scientists believe that psi is real, but like a hot potato it's too uncomfortable to handle. Others believe that psi is a childish novelty unworthy of serious attention.

The majority who believe that psi is real are forced to confront the problem of "forbidden knowledge," taboo topics that restrict the conduct, funding, and publication of certain ideas. An article on this issue in the journal *Science* in 2005 described the results of a survey on forbidden knowledge from scientists at prestigious academic departments in the United States. It found that most felt that "informal constraints" limited what they could study. These constraints included concerns over what they thought the news media, journal editors, activists or peers might think of their interests. Because of such social and political pressures, scientists shy away from controversial topics. As one respondent in the survey put it, "I would like to lunatic-proof my life as much as possible."[2]

This is the state of affairs for research on ordinary topics, so you can imagine the situation for psi research. Traditional sources of funding hardly ever consider touching hot potatoes, and as a result there are fewer than 50 conventionally trained doctoral-level scientists around the world engaged in full-time psi research. A common feature among members of this group is that they're not intimidated by orthodoxy. As one of those card-carrying iconoclasts, I'm often asked why I became interested in psi research, and why I persist in such an apparently quixotic quest. To answer those questions, I'll briefly review my background.

A PERSONAL VIEW

I've been consumed with the question of personal existence for as long as I can remember. In the first grade, when morning recess period began the other children scampered out to the playground to cheerfully stomp on each other. But I hung behind to pepper my teacher, Miss Platt, with existential questions. Elementary school teachers are not paid nearly enough to indulge annoying kids troubled by philosophical uncertainties, especially before lunchtime.

I was fascinated with such questions as, Why are we here? Is this all we're capable of? Does life have any real purpose, or is all this emphasis on arithmetic and spelling just a distraction to avert our attention from more important questions, like the futility of existence? One day, some kids were shooting spitballs in class and creating a commotion. As I watched the uprising unfold, I remember thinking, "What's wrong with those kids? They're acting like children!" The instant this thought came to mind, I was struck with a peculiar moment of mental clarity. These couldn't be *my* thoughts. These were the exasperated thoughts of an adult charged with supervising gangs of misbehaving, preadolescent primates. But I was one of those mischievous monkeys, so what was I doing thinking such thoughts? Like a thunderbolt, I realized that "me" and my thoughts weren't necessarily the same thing. In retrospect, perhaps I was empathizing with our poor teacher, whose face was slowly congealing into the permanently anguished expression made famous in Edvard Munch's painting, *The Scream*. This episode sparked similar incidents of acute self-awareness, and it led me to ponder questions like, What is the "I" that's watching my thoughts? And, Who's asking *that* question?

The curiosity wasn't due to existential angst. I had a happy childhood, and I was raised in a blithely agnostic, artistic family. My interests were undoubtedly inherent; a friend joked that I must have been born with an extra set of "why" chromosomes.

Whatever the cause, my interests in the human mind were further catalyzed by my first career. I started playing the violin at age 5, and before I knew it I had spent the next 20 years performing as a classical soloist and in orchestras and quartets.

During those formative years, my parents and violin teachers teased me with a certain phrase. After I'd finish practicing a difficult piece, they'd say, "That was good, but you're not living up to your potential." This mantra was undoubtedly meant to inspire me to work harder. But its actual effect was to cause me to wonder, with growing intensity over the years, what *is* my potential? How would I know when I've achieved it? What *are* the farthest reaches of the human mind?

My absorption with such questions eventually steered me away from a career in music. Instead I became attracted to the question of human capacities and potentials. Today, after spending the majority of my career investigating this question, I have yet to find where human potential ends. The more I look into it, the more I discover how much is left to learn. I've come to agree with Willis Harman, president of the Institute of Noetic Sciences from 1977 to 1997, who wrote extensively on these issues. Harman succinctly summarized the situation as: "Perhaps the only limits to the human mind are those we believe in."

Of the many interesting topics associated with human potential, one quickly caught my attention—psychic experiences. I became interested in psi around age 10, after I discovered that wonderful section of the public library that housed the fairy tales, mythology, parables, and science fiction. In those fictional realms, it was taken for granted that the mind had exceptional powers and capabilities, and it all seemed perfectly reasonable to my youthful intuition. Around this time, I also discovered that I had an affinity for science and math. But it was clear that science and fairy tales shared only one commonality—creative imagination. Beyond that, I didn't see how they could possibly have any connection.

Still, I was struck by the way that psychic powers in yogic lore, the *siddhis,* were described in such matter-of-fact tones in ancient texts, like in Patanjali's *Yoga Sutras.* Such abilities were not portrayed as supernatural fairy tales, but as pragmatic, ho-hum consequences of practicing meditation. Such claims far exceeded the bounds of science, but the authors of those books seemed thoughtful and intelligent, and they appeared to be as adept at investigating "inner space" as Western scientists had become at investigating "outer space." Surely, I thought, such claims were just children's stories, mere imaginative fantasies of prescientific peoples. I found it natural to adopt a skeptical stance, as my interests were not motivated by frequent psi experiences, but rather by strong curiosity and a natural disposition for empathy.

One day after I had pestered a kindly librarian with one too many questions, she introduced me to books describing scientific investigations of psychic phenomena. I was hooked. I discovered that there *was* a link between psychic abilities and science—these abilities can be tested in the laboratory.

I soon learned that for every two books I read that presented scientific evidence for psi, I found a third that countered it. I'd read a skeptical book and find myself feeling cynical about the varieties of human stupidity, and then I'd read a book by a scientist who had actually conducted experiments and find myself becoming excited about the prospects of exploring the frontiers of the human mind. Both sides of the controversy seemed sensible enough; both argued their side convincingly. But after reading all of these books I noticed that the debate followed a predictable pattern: One side presented experimental evidence that something interesting was going on; the other argued that the evidence wasn't good enough to be taken seriously. Some skeptics pushed doubt to extremes and insisted that positive evidence was *always* due to mistakes or intentional fraud.

As I saw it, within this dialectic one side was struggling to understand the depths of inner space by probing Nature with clever questions. The other was trying to maintain the status quo through passionate, and sometimes vicious, denial. The former were willing to take risks to advance knowledge, the latter were naysayers interested mainly in defending dogma. I found the explorers far more interesting than the skeptics, and I was impressed to learn that some of the greatest minds in modern times, people like Nobel laureate physicist Wolfgang Pauli and psychoanalyst Carl Jung, were deeply interested in psi.[3]

About fifteen years after I started following the literature on psi research, I had earned a master's degree in electrical engineering and a doctorate in psychology from the University of Illinois at Urbana-Champaign. Besides psi, I was interested in cybernetics and artificial intelligence. My electrical engineering thesis advisor was Heinz von Foerster, a pioneer in the foundations of cybernetics. One of von Foerster's lasting contributions was clarifying the role of self-reference in complex domains, including such perplexing conditions as when the observer observes himself. Cybernetics gave birth to topics known today as self-organizing systems and chaos theory, and I continue to be interested in those disciplines as I believe that self-reference and psi are deeply related.[4] My graduate work involved designing computer models of cognition and applications of artificial intelligence. My advisor was Andrew Ortony, known for his research on cognition and emotion. One of the members of my doctoral committee was John Bardeen, one of only four scientists to be awarded two Nobel Prizes.

Throughout my university years I dabbled with simple psi experiments, but after joining the staff at AT&T's Bell Laboratories,[5] I began to attend the annual conferences of the Parapsychological Association (PA). The PA is the international professional organization for scientists and scholars interested in psi phenomena; it's been an affiliate of the American Association for the Advancement of Science (AAAS) since 1969.[6] Later I worked at SRI International (formerly known as Stanford Research Institute) on a classified program investigating psi phenomena for the US government.[7] Then I held appointments at Princeton University, the University of Edinburgh, the University of Nevada, and two research labs in Silicon Valley. In 2001, I joined the research staff of the Institute of Noetic Sciences.

At work, I spend most of my time analyzing data, writing articles, and preparing or conducting experiments in the laboratory. Occasionally I can be caught gazing out the window trying to make sense of the intriguing effects we observe in our experiments, or musing over a puzzling problem while playing bluegrass on my five-string banjo. During one of those flights of fancy, I found myself pondering a restlessness brewing in science.

SHAKEN ASSUMPTIONS

Unexpected discoveries across many disciplines are shaking previously held assumptions. One commonality is that observations previously thought to be meaningless anomalies are being reconsidered. In the process, new revelations are surfacing about the nature of reality.

- Cosmologists have learned that we might have accidentally overlooked 96% of the universe. The missing majority of the universe has been dubbed "dark" energy and matter. We know next to nothing about it, and it's spawning whole new concepts about the structure and evolution of the universe.[8] As theories of cosmology are being reconsidered, new light is dawning on astronomical anomalies first observed decades ago.[9]
- Molecular biologists, who recently regarded large segments of the genome as "junk DNA" because no one knew what it was good for, have been astonished to find strong commonalities among DNA base-pairs in humans, chickens, dogs, and fish. It appears that some aspects of DNA have been ultraconserved for hundreds of millions

of years, and that previous assumptions about what is important in DNA were wrong.[10]

- For a century, neuroscientists believed that neurons in the brain do not regenerate, that once there is a brain injury or as neurons die in the course of aging, normal mental functioning inevitably deteriorates. Now we've learned that the dogma was wrong—brain neurons do regenerate. The plasticity of the brain is much greater than previously expected.[11] This helps shed new light on previous observations that were ignored because they didn't make any sense. In a case study reported in 1980 in the journal *Science,* during a routine examination for a minor medical ailment, a student at Sheffield University in Great Britain was found to have virtually no brain.[12] But that didn't stop him from enjoying an IQ of 126 and graduating with first-class honors in mathematics.

- A 2004 U.S. government–sponsored review has resurrected interest in "cold fusion" after a 15-year hiatus in the scientific deep freeze.[13] Successful replications of cold fusion phenomena from laboratories around the world continue to suggest that unexpected effects do occur. Understanding what is going on may require a new scientific discipline that straddles nuclear physics and electrochemistry.

- Cosmologists working on mathematical models of black holes have been surprised to find that the entire universe can be described as a type of hologram, or as interference patterns in space and time. As Stanford University physicist Raphael Bousso wrote in *Scientific American,* "The amazing thing is that the holographic principle works for all areas in all space times. We have this amazing pattern there, which is far more general than the black hole picture we started from. And we have no idea why this works."[14]

ENTANGLEMENT

But perhaps the most significant discovery is *entanglement,* a prediction of quantum theory that Einstein couldn't quite believe, calling it "spooky action at a distance." Erwin Schrödinger, one of the founders of quantum theory, used the word entanglement to refer to connections between separated particles that persisted regardless of distance. These connections are instantaneous, operating "outside" the usual flow of time.[15] They imply that at very deep levels, the separations that we see between ordinary, isolated objects are, in a sense, illusions created by our limited perceptions. The bottom line is that physical reality is connected in ways we're just beginning to understand.

Entanglement was predicted based upon the mathematics of quantum theory. It was originally thought to be so fragile that, in the estimation of a prominent physicist, "anything, even the passage of a cosmic ray in the next room, would disrupt the [quantum] correlations enough to destroy the effect."[16] Today we know that entanglement is not just an abstract theoretical concept, nor is it a quantum hiccup that only appears for infinitesimal instants within the atomic realm. It has been repeatedly demonstrated as fact in physics laboratories around the world since 1972. As research accelerates on this surprising characteristic of nature, entangled connections are proving to be more pervasive and robust than anyone had previously imagined.[17] A review of developments on entanglement research in March 2004 by *New Scientist* writer Michael Brooks concluded that "Physicists now believe that entanglement between particles exists everywhere, all the time, and have recently found shocking evidence that it affects the wider, 'macroscopic' world that we inhabit."[18]

A FANTASTIC SCENARIO

I believe that entanglement suggests a scenario that may ultimately lead to a vastly improved understanding of psi. The scenario begins with the exploding use of digital information systems in every realm of modern life. The need to keep that information secure has placed massive pressure on the computing and communication industries, and it has generated a need for computers that can process information thousands of times faster than today's fastest supercomputers. One possible solution is quantum computing. It has been estimated that a single quantum computer could theoretically perform more computations than would be possible for a classical computer the size of the entire universe.[19] Such electrifying pronouncements have attracted substantial funding and as a result, research in quantum communication and information processing is rapidly advancing.

Articles reporting new developments in entanglement theory and applications now appear regularly in scientific journals. Demonstrations of entanglement initially relied on extremely sensitive measurements in exotic conditions like extreme cold or incredibly short periods of time, but now researchers are reporting increasingly complex forms of entanglement that are lasting for much longer periods of time, and at higher temperatures. For practical uses like quantum computers, proposals like "entanglement purification" and "coherence repeaters," which are ways of extending the special quantum states required to sustain entanglement, are likely to be further refined to allow increasingly large objects to remain entangled at room temperature and for indefinite lengths of time.[20]

Physicists have been able to entangle ensembles of trillions of atoms in gaseous form, and entanglement has been demonstrated among the atoms of relatively large chunks (centimeter-square) of salt.[21] Entangled photons shot through sheets of metal have been shown to remain entangled after punching through to the other side.[22] Photons also remain entangled after being sent through 50 kilometers of optical fiber, and while being transmitted through the open atmosphere. Clusters of four entangled photons have been demonstrated to make quantum computing significantly easier to accomplish than it was previously imagined.[23] And organic molecules, like tetraphenylporphyrin $(C_{44}H_{30}N_4)$, have been successfully entangled.[24]

While practical difficulties must be overcome before entanglement is demonstrated in viruses, proteins, and living systems, there's no theoretical limit to how large an entangled object can be. Of course, physicists are quick to point out that when carefully prepared atomic-sized objects interact with the environment, by say colliding with air molecules or passing through electromagnetic fields, they become entangled with those objects. Those interactions tend to quickly smooth out the special state of quantum "coherence" in which simple forms of entanglement can be most easily observed. This loss of coherence, appropriately called decoherence, is (among other reasons) why we perceive everyday objects as separate and not as blurred together. But decoherence doesn't magically make quantum effects vanish. We're still thoroughly permeated by entangled particles. The question posed here is whether these deeply entangled states are meaningfully related to human experience, and if so, are they also related to psi? I propose that the answers are yes and yes, as we'll see.

One reason is that some scientists now believe that bioentanglement—quantum connections within and among living systems—will be useful in explaining the holistic properties of life itself. Numerous scientists, including Nobel laureate physicist Brian Josephson, have also proposed that biological systems might find ways of *using* entanglement in novel ways.[25] In 2005, physicist Johann Summhammer, from the Vienna University of Technology, proposed that because entanglement is everywhere in nature, it's conceivable that evolution has taken advantage of it. In particular, he proposed that

> Entanglement would lead to a Darwinian advantage: Entanglement could coordinate biochemical reactions in differ-

ent parts of a cell, or in different parts of an organ. It could allow correlated firings of distant neurons. And . . . it could coordinate the behavior of members of a species, because it is independent of distance and requires no physical link. It is also conceivable that entanglement correlates processes between members of different species, and even between living systems and the inanimate world.[26]

Physicists have even speculated that entanglement extends to everything in the universe, because as far as we know, all energy and all matter emerged out of a single, primordial Big Bang. And thus everything came out of the chute already entangled. Some further speculate that empty space, the quantum vacuum itself, may be filled with entangled particles.[27] Such proposals suggest that despite everyday appearances, we might be living within a holistic, deeply interconnected reality.[28] To be clear, these speculations are being proposed by traditional physicists, not by starry-eyed new agers or mystics.

THE FUTURE

In the near future, when the concept of entanglement is better understood, I expect that someone will get a bright idea and ask, "I wonder what would happen if two human beings became entangled? Perhaps they'd show correlated behavior at a distance too, just like entangled atomic matter does." Case studies of identical twins will be used to justify this speculation. For example, consider the true case where twin boys raised separately were independently named "Jim" by their adoptive parents. Each Jim married a woman named Betty, divorced her, then married a woman named Linda. Both Jims were firemen, and each built a circular white bench around a tree in his backyard.[29] Could such coincidences arise from common genes that programmed Betty tendencies, Linda tendencies, and firemen tendencies? Or does it reflect "entangled Jims"?

Intrigued by such stories and by demonstrations of bioentanglement, an enterprising scientist will conduct an experiment. She'll isolate two identical twins in dark, soundproof and electromagnetically shielded chambers. She'll ask them to keep each other in mind while at random times she'll flash a bright light at one of them. Each of those light flashes will generate a predictable response in that twin's brain. After confirming the presence of those responses, she'll examine the brain activity of the *other*, nonstimulated twin, to see if there's a corresponding response at the same time. This electroencephalograph or "EEG correlation" experiment will successfully demonstrate a positive correlation between the two brains, and it will be widely hailed as a breakthrough of stunning proportions.

Then someone will quietly ask, "I wonder what it *feels like* when my brain is entangled with another brain." And then the panoply of psi phenomena will be rediscovered for the umpteenth time. But this time, for the first time, it will be accompanied by a solid theoretical foundation.

How long will we have to wait before this fanciful scenario unfolds? No time at all. The "entangled brains" experiments have already been performed over a dozen times over the past 40 years by independent groups.[30] *And they work.*

One of the first such experiments was published in 1965 in the journal *Science*. That study reported that the EEGs of pairs of separated identical twins (two such pairs out of 15 pairs tested) displayed unexpected correspondences. When one twin was asked to close his or her eyes, which causes the brain's alpha rhythms to increase, the distant twin's alpha rhythms were also found to increase.[31] The same effect was not observed in unrelated pairs of people.

Today, positive results in these EEG correlation experiments continue to be reported. A notable advance was published in 2003 by Leanna Standish and her colleagues at Bastyr University. This was an experiment not using EEGs but the brain scanning technology known as functional magnetic resonance imaging (fMRI). Standish found in one selected pair of participants that the visual cortex of the "receiving" person's brain became activated when her distant partner was exposed to a flashing light. This outcome was consistent with the results of the EEG correlation studies, but it also located precisely *where* in the brain the effect occurs. In 2004, psychophysiologist Jiří Wackermann published a review of this class of experiments in *Mind and Matter,* a new scholarly journal devoted to interdisciplinary research on the mind-matter interaction problem. Wackermann concluded that there appears to be a real, repeatable effect, and it's encouraging that with increasingly sophisticated experimental designs the effects continue to be observed by independent investigators.[32]

ENTANGLED MINDS

This book suggests that we take seriously the possibility that our minds are physically entangled with the universe, and that quantum theory is relevant to understanding psi. That said, we should avoid jumping to premature conclusions. I'm not claiming that quantum entanglement magically explains all things spooky. Rather, I propose that the fabric of reality is comprised of "entangled threads" that are consistent with the core of psi experience. Of course, human experience is far more than a col-

lection of threads. Our bodies are tapestries built from countless variations of the fabric of reality. And our subjective experiences (to stretch a metaphor) are quilts made from tapestries that are stitched together in myriad, delightful ways. Understanding the nature of this quilt, and its relationship to psi, will take more than identifying the nature of the threads that weave the fabric of reality. But it's an important first step. And it provides a new perspective from which to pose questions that may lead to unexpected answers about psi.

CHAPTER 2

NAKED PSi

How do you know but ev'ry Bird that cuts the airy way,
Is an immense world of delight, clos'd by your senses five?

–William Blake

Once upon a time, in a sleepy country town far, far away, there was an introspective young boy named Hans.[1] Hans was more interested in his grandfather's poetry and in stargazing than in becoming a doctor like his father. After finishing high school, he decided to attend the university in the city, aspiring to become an astronomer. But the pace of big city life disagreed with his quiet ways, and after a short time he left school. It was a time of peace, so he decided to enlist for a year of service in the cavalry, looking forward to a year of riding horses and enjoying the outdoors in relative serenity.

One morning, while he was on horseback during a training exercise, his horse suddenly reared. Hans was tossed into the air and he landed hard on the road directly in the path of a fast-approaching, horse-drawn cannon. He realized with horror that he was about to be crushed, but miraculously, the driver of the artillery battery managed to stop the horses just in time. The accident left Hans thoroughly shaken but without serious injury.

At that very moment, many miles away in his family's

home, Hans's older sister was suddenly overwhelmed with an ominous certainty that something bad had happened to Hans. She anxiously insisted that their father contact him, and so he did via a telegram.

That evening, when Hans received the telegram, he was initially concerned, as he had never before received a telegram from his father. Then, upon reading his sister's urgent concern about his well-being, he knew that his feelings of intense fear earlier in the day had somehow reached his sister. Many years later, Hans wrote, "This is a case of spontaneous telepathy in which at a time of mortal danger, and as I contemplated certain death, I transmitted my thoughts, while my sister, who was particularly close to me, acted as the receiver."[2]

This experience profoundly transformed Hans's interests from the depths of outer space to the depths of the human psyche. After he finished his military service, he immediately returned to the university and focused on learning medicine, determined to understand how "psychic energy," as he called it, could carry a telepathic message to his sister a hundred miles away.

After many years of concentrated effort, working mostly in solitude in his laboratory at the university, Hans finally developed a method of recording human brain waves. For a time they were called "Berger rhythms," after Hans's last name. Now we call these signals an electroencephalogram, or EEG. With this invention he established for the first time that electrical activity of the human brain was correlated with different subjective states of mind. But Hans didn't forget his original passion; he also carried out an experimental program involving 200 subjects, each of whom was tested for telepathy while in a hypnotic trance.

Hans's driving passion to understand psychic energy did not succeed in explaining his sister's telepathic experience, but it did establish the foundations of modern neuroscience. We are indebted to Hans not only for his development of the EEG, but also for revealing the basic brain mechanisms used in medical imaging devices like positron emission tomography (PET) and functional magnetic resonance imaging (fMRI).[3]

Tragically, as is all too common when it comes to scientific breakthroughs, Hans didn't live to enjoy his well-earned recognition. Most of his scientific peers around the world believed that his recordings were due to some sort of electrical or mechanical artifact. Even his own colleagues considered him a naïve amateur and a suspect loner. After a long illness, despondent, and suffering from a painful skin infection, he committed suicide in 1941.

This is the true story of Hans Berger, German psychiatrist and father of the EEG. His invention sparked the rapid development of ever-more sophisticated ways of measuring brain activity. There's little doubt that Hans would have been deeply satisfied if he could have known that a quarter-century after his death his discovery would spark a new chapter in the quest to understand the "psychic energy" he was seeking most of his life.[4]

As with the untold history of the EEG, most textbooks present well-mannered, sanitized origins of the major scientific discoveries. Naturally, the true origins of ideas are far more circuitous and perplexing. Most neuroscientists today don't know that their discipline was inspired by a telepathic experience, or that the functions of the cerebral cortex, corpus callosum, and corpus striatum were all accurately described two hundred years before the rise of modern neuroscience.[5] Most medical scientists don't realize that the gold standard "randomized controlled trial" design used in clinical research was initially developed to investigate psychic phenomena. The same can be said for key developments in clinical psychology, mind/body medicine, psychophysiology, and experimental psychology. Even the discovery of isotopes, an advancement that helped pave the way to the atomic bomb, can be traced to a case of clairvoyance.

Modern science itself might have been spawned in a series of feverish dreams on the night of November 10, 1619, by a 24-year-old Frenchman named René Descartes. He had three dreams that evening, involving terrifying phantoms, whirlwinds, fiery sparks, and books of symbolic wisdom. Those dreams are said to have inspired Descartes to found the principles of rational empiricism.[6] Coincidently, that same evening, November 10, was St. Martin's Eve. A traditional ceremony performed on St. Martin's Eve is a procession with lamps, used to symbolize the bringing of spiritual light into the darkness. As Descartes dreamt of banishing the darkness of ignorance, all over Europe rituals were taking place, seeking a similar goal.

WHEN FICTION BECOMES FACT

The science fiction author Philip K. Dick had a uniquely imaginative mind. He published 112 short stories and over 30 novels.

Many of his stories were turned into popular science fiction movies, including the cult classic *Blade Runner,* and later *Total Recall* and *Minority Report.* His interest in the nature of reality and time was motivated by many unusual personal experiences associated with his stories. One such episode he describes is as follows:

> In 1970 I wrote a novel called *Flow My Tears, the Policeman Said.* One of the characters is a nineteen-year-old girl named Kathy. Her husband's name is Jack. Kathy appears to work for the criminal underground, but later, as we read deeper into the novel, we discover that actually she is working for the police. She has a relationship going on with a police inspector. The character is pure fiction. Or at least I thought it was.
>
> Anyhow, on Christmas Day of 1970, I met a girl named Kathy—this was after I had finished the novel, you understand. She was nineteen years old. Her boyfriend was named Jack. I soon learned that Kathy was a drug dealer. I spent months trying to get her to give up dealing drugs; I kept warning her again and again that she would get caught. Then, one evening as we were entering a restaurant together, Kathy stopped short and said, "I can't go in." Seated in the restaurant was a police inspector whom I knew. "I have to tell you the truth," Kathy said. "I have a relationship with him."
>
> Certainly, these are odd coincidences. Perhaps I have precognition.[7]

PREMONITIONS OF 9/11

The following is an excerpt of a premonition involving the collapse of the World Trade Towers in New York City during the terrorist attacks of September 11, 2001. Physician Betsy MacGregor and her husband, Charles, were on a plane, flying to their home on an island in Puget Sound near Seattle after visiting friends in New York City. The date was September 10, 2001; it was midnight on the airplane. Dr. MacGregor writes:

> There were relatively few people on the flight, and many seats were empty . . . . Spotting an empty row across the aisle, I decided to move there and stretch out. I arranged myself fairly comfortably in the new row, grateful to be able to lie down. I was still quite tired and expected to fall back asleep promptly. But something didn't feel right. I grabbed a couple of extra pillows to soften the lumps in the seats, but I couldn't seem to relax . . . I emptied my mind of all thoughts, focused my attention on the muted roar of the engines, and lay very still—more awake than ever.
>
> In the beginning it was almost imperceptible, the strange feeling that started to come over me. It began with an awareness of how absolutely still my body was. I wondered vaguely why it was so perfectly motionless and felt a growing urge to move it. But when I sent out the intention to move, to my surprise my limbs did not respond. I wondered if I was asleep and having a bizarre dream in which I seemed to be awake but wasn't. The more I tried to move, however, the more I detected a kind of resistance. Something hard and unyielding surrounded my body, immobilizing it. Yes, I felt it clearly now, I was completely encased and held fast in concrete . . . .
>
> The feeling of being imprisoned in concrete intensified—with it now was a sense of dread. I could not turn my head or move my arms or legs or expand my lungs with a deep breath of air. I was hopelessly trapped and on the verge of claustrophobic terror.
>
> And then the pain began. Faint at first, it rapidly grew stronger until it filled my whole body. For the concrete was moving. From all sides it was pressing in on me, tighter and

tighter, squeezing me with unbearable force. My body was about to be crushed.

A voice in me screamed out: *No! Not possible! How can this be?* For a split second my mind spun around wildly, refusing to believe, looking for a way out. But it was absolutely, perfectly clear: there was no escape. There was nothing to hold on to, nowhere to run to. In another instant my life would be over. I saw that. *I saw death before me* . . . .

What was that all about? I had no idea how much time had passed; it could have been minutes, or it could have been hours. I lay there for a long while, completely mystified. . . . It was a long trip from the Seattle airport to the ferry and across the Puget Sound to the island where Charles and I live. We were totally exhausted when, a little before 6:00 a.m. Pacific time, we finally arrived home. As the pale light of dawn was spreading across the eastern sky, we headed upstairs and tumbled gratefully into bed. Three thousand miles away, the north tower of the World Trade Center was bursting into flames. Shortly thereafter a second plane roared into the south tower. As Charles and I slept, stunned New Yorkers—many dear friends of ours among them—gaped in horror and disbelief as first one tower, and then the other, crumbled into dust. Thousands of lives ended that morning in the crush of concrete.[8]

Here is a similar experience, reported a few weeks before the 9/11 terrorist attacks by a woman named Marie. This is just one story from a compilation of 14,000 cases of spontaneous psi experiences collected by the Rhine Research Center over many years.[9]

When we exited the city, my husband was driving. I was sitting next to him in the front. I was just trying to close my eyes to relax for a minute. Then he told me, 'Well, when we come around the bend up ahead, you should get a good view of the Pentagon because our road goes right by it.' It was one of the things we had said we wanted to do when we visited Washington. So I opened my eyes to look, and when I looked to the right, there it was. But it had huge billows of thick, black smoke pouring out of it, just huge clouds of smoke. I didn't see fire, I saw smoke, like a bomb had gone off, billows and billows of black smoke going up in the sky.

I yelled out and slammed my hands on the dashboard. My poor husband didn't know what was happening. I mean, I really screamed out loud. His first thought was that we were going to be in an accident, and I was warning him he was going to hit someone. But it was pretty open space on the highway, and nobody was cutting in front of us or anything at that moment.

I truly felt like we were in danger, even though we were actually on the highway and a couple of miles away from the Pentagon. I thought it was on fire. My husband said the Pentagon was not on fire, and then I finally realized that in fact it wasn't. And as fast as it had started, it stopped. It had all happened in a few seconds.

Many similar forebodings of 9/11 have surfaced. Are they true premonitions, poignant coincidences, or due to psychological frailties like selective memory or wishful thinking? Given the billions of dreams experienced nightly by people around the world, we would expect to hear about occasional "miraculous" coincidences every so often. How then can we tell if a premonition is real or illusory? And why, given the horrific circumstances of 9/11, or the colossal tsunami tragedy of December 2004, aren't more such premonitions reported? From a basic science point of view, what we'd like to know is whether such premonitions are possible even *in principle*.

UNCONSCIOUS PREMONITIONS OF 9/11?

Most of this book is concerned with the "in principle" question. Before we begin that expedition, let's pause to consider a new approach to studying the question of premonitions. In September 2000, I designed a suite of Web-based games located at www.GotPsi.org and hosted by the Boundary Institute, a Silicon Valley thinktank I cofounded with computer scientist Richard Shoup. This Web site allows users to test their psychic abilities online. All of the data contributed there are recorded for research purposes, and as of late 2005 the database consisted of over 60 million individual trials contributed by almost a quarter million people worldwide.

One of the tests on that site assesses precognitive ability. It tests how well a user can describe a photo that the computer will randomly select *after* the user enters a description of the photo. The description can be entered in the form of words, or by checking boxes indicating whether the user thinks the photo will be of an indoors or outdoors scene, will involve people, will have water present, and so on. Because this test asks people to imagine a visual scene they are about to see, I thought it might be interesting to investigate whether premonitions of 9/11 might have spontaneously intruded into their attempts to describe an image the computer would soon display. So I looked at the words that people used to describe their imagery from September 9 through the morning of September 11, 2001. This included a set of about 900 trials and just over 2,500 words.

On Sunday, September 9, 2001, between 8:48 and 8:57 a.m. Eastern Daylight Time, a user nicknamed *sean* wrote the following impressions in a series of three successive trials:

> airliner (seen from left-rear) against stormy cloud backdrop, flashes of streaky cloud, ovoids, two persons
> firstly a dragonfly? then a log [or] branch suggestive of

Everglades, then a fast dynamic scene of falling between two
tall buildings, past checkered patterns of windows

first tall structure like an industrial chimney, then flashes
of rounded crenulated form—peacock-like headdress of American Indian woman? then surface like volcanic ash plume or
cauliflower

Sean's precognitive descriptions failed to match the photos
subsequently selected at random by the computer. But they do
provide a rather startling impressionistic sense of the chaos associated with the events of 9/11 in New York City. The next
day, September 10, 2001, starting at 5:00 p.m. Eastern Time,
user *shakey* wrote these words in two successive trials:

it is of something falling; it will be a chaotic scene

Again these were poor descriptions of the targets, but meaningful in the context of 9/11. A half hour later, a different user,
nicknamed *justatest,* wrote in four successive trials:

intense . . . too hot to handle; blasting; is the coast clear?;
they were checking the coast!!!

The following morning, Tuesday, September 11, 2001,
about an hour before the first airplane crashed into the World
Trade Center tower, user *xixi* wrote the following words in a series of 11 trials:

White House; gone in the blink of an eye; scald; man's
folly; band red; surging; palace; not easily conned; US power
base; flexing muscles; surprise.

Are these genuine premonitions of 9/11? The ideas suggested by these words were unusual in the context of this online
precognition experiment, as most of the photos used in the test

are of benign landscapes, people, animals, and other pleasant scenes with neutral content. Still, this is just a handful of potentially interesting matches out of 900 trials, and arguments based solely on subjective assessments have little currency in science. So I devised a way to judge whether the words used in this precognition test prior to 9/11 were in fact unusual.

MASS PREMONITIONS OF 9/11?

I should emphasize that the following is speculative and not representative of the controlled laboratory work we'll discuss later. Nevertheless, I present it because it illustrates a way that web-based experiments are beginning to offer new ways of studying *collective* psi effects. With that caveat in mind, I first examined the data from all the online precognition trials contributed from September 2, 2000, through June 30, 2003.[10] There were 428,000 trials contributed by about 25,000 people. From those data, I selected only trials that included word descriptions; this included 256,000 trials and 841,000 words.

For each of those trials, I matched the words entered against a set of nine concepts that captured the chaotic context associated with 9/11: *airplane, falling, explode, fire, attack, terror, disaster, pentagon,* and *smoke.* The idea was to see how closely the words provided by hundreds of users each day matched these concepts. Counting only exact word matches wouldn't be fair because someone might have used a synonym or an associated word that an exact wordmatch would overlook. So I developed a computer-based *concept* matching technique to create a daily terrorism ideation score.[11]

This analysis showed, to my surprise, that on 9/11 the curve dropped to its lowest point in 3 years of collecting data (Figure 2–1). Rather than increase in value, as might be predicted if lots of people were suddenly having spontaneous premonitions of disaster, and inadvertently reporting those impressions in this online test, the scores significantly *dropped* as 9/11 approached.

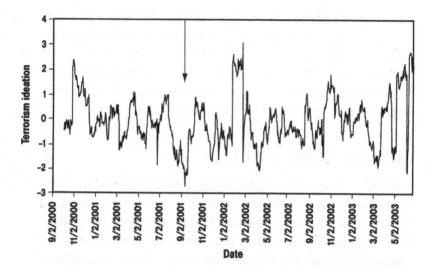

Figure 2–1. Fluctuations in daily terrorism ideation scores from September 2000 through June 2003. The arrow points to September 11, 2001. The dip at 9/11 suggests that participants in an online remote viewing test were actively avoiding terrorism-related concepts just before 9/11.

A statistical test compared to similarly constructed, randomly scrambled datasets showed that the odds against the chance of obtaining a terrorism ideation score as low as the observed minimum, and falling on 9/11 as observed, was 3,300 to 1.[12] Thus, the data didn't indicate that premonitions intruded into people's thoughts just prior to 9/11. Rather it suggests that, on average, such thoughts were significantly *avoided*.

If this isn't a coincidence, then what might cause such an effect? One possibility is that in the days before 9/11 many people begin to unconsciously sense trouble brewing, but there was no context for those feelings so they were repressed. Repression is an unconscious psychological mechanism we use to actively avoid disturbing emotions or images. No one wants to walk around with troubling images of disasters rattling around in their heads, so repression is expected. Only the rare individual can avoid personally identifying with negative thoughts without repressing them, and fewer still are willing to publicly admit such thoughts. This may be why verified premonitions of major disasters that are recorded before the fact are relatively rare.

MORE PREMONITIONS OF 9/11?

If the repression idea has any merit, then we might expect it to show up in other psi performance tests. As it turns out, I was also examining data from another online test at www.GotPsi.org—a card guessing test. In that game, you see five cards on the screen and are asked to select one that you think the computer will select later. You make your selection, then the computer randomly selects a card and displays it. By chance you'd expect over the long run to correctly guess 1 in 5, or 20% of the cards. From August 2000 through June 2004, this online test collected 17 million trials. For each of those days we can form a score reflecting the performance, contributed by hundreds of users per day, compared to chance expectation.[13]

The results showed a huge drop in performance observed prior to 9/11 (Figure 2–2). The odds against chance of seeing a drop that deep, and as close or closer to 9/11 as observed, is 2,700 to 1.[14] This means that the users were actively *avoiding* hitting the correct card just prior to 9/11.

This outcome is consistent with the possible repression effect that we observed in the precognition test. Together they suggest that days before 9/11 many people may have been unconsciously avoiding their psi impressions to suppress awareness of a looming disaster. While this is purely speculative, the likelihood that two independent online tests would both display strong negative tendencies just prior to the same meaningful date is associated with odds against chance of 1.8 million to 1.[15] That seems to offset the possibility that we're dealing with a mere coincidence.

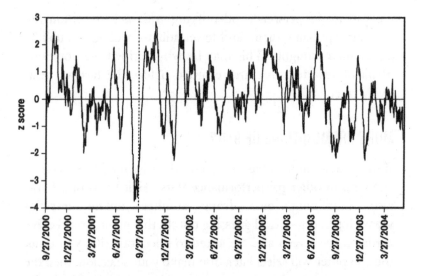

Figure 2–2. Fluctuations in online ESP card test performance on a daily basis from September 2000 through June 2004. The pronounced dip just before 9/11, indicated by the dashed vertical line, suggests that participants in this test avoided selecting the correct card.

But are spontaneous reports of psi experiences due to coincidence? Perhaps widespread belief in psi merely reflects ignorance of scientific principles and methods. Or maybe it's due to sifting through mounds of data in the belief that something mysterious must be hiding there. In exploring the unknown, we must be prepared for all possibilities. So let's continue by examining the issue of belief.

CHAPTER 3
WHO BELIEVES?

When a belief is widely held in the face of overwhelming evidence to the contrary, we call it a superstition. By that criterion, the most egregious superstition of modern times, perhaps of all time, is the "scientific" belief in the non-existence of psi.

—Thomas Etter[1]

The ideal in science is to allow our experiences, in the form of formal observations and measurements, to rationally shape our beliefs. We do this through controlled experiments. In practice, we can't personally experience everything, so we're obliged to place our faith in what others report. When faith collides with experiments, disagreements invariably arise. We usually think of this conflict in terms of religion vs. science. But sometimes disagreements arise because *scientific faith* clashes with repeated *human experiences*. When this happens, emotions trump reason. Let's examine one of these disputes.

If we are to believe the assertions of the scientific mainstream, scientists view the general public as stupid, tobacco-spitting hillbillies who "ain't got no book-larning." Why? Because the unwashed masses believe in things that science regards as beyond rational discourse, or reeking of superstition, or as just plain impossible.

That stereotype is a bit harsh. And I'm sure that tobacco aficionados from the Appalachian Mountains do not appreciate such sentiments. But according to a key document published by the U.S. government's National Science Foundation (NSF), you'll find that this statement isn't all that far off the mark. The NSF believes that the majority of the general population is stupid because they believe in psi and other "pseudosciences."[2] In this chapter we'll investigate who's likely to be closer to the truth, the NSF or the hillbillies.

Let's consider two flavors of stupidity: Just Plain Stupid and Mentally Deficient. For the sake of science, let's dignify Just Plain Stupid into something more official sounding by calling it *the ignorance hypothesis.* This proposes that people believe in the paranormal because they're uneducated. The assumption is that if only people would pay more attention to what science teaches about the way the world works, then they'd stop believing in delusions like telepathy. Everyone knows, so this hypothesis goes, that concepts like psi violate basic scientific laws, thus anyone who is unaware of such elementary laws must be ignorant and is therefore likely to believe in anything or anyone. That in turn threatens the fabric of a civilized, rational society, and must be squashed. A testable prediction of this hypothesis is that *lower* levels of formal education ought to be associated with *higher* levels of belief in psi.

The second form of stupidity we'll call the *mental deficiency hypothesis.* It asserts that superstitious beliefs arise in some people because they're dim-witted or mentally ill. Like the ignorance hypothesis, the mental deficiency hypothesis is taken for granted by some within the medical orthodoxy. For example, in the 1994 edition of the American Psychiatric Association's "Diagnostic and Statistical Manual of Mental Disorders" (called the DSM-IV), a portion of the description for *schizotypal personality disorder* is as follows:

A pervasive pattern of social and interpersonal deficits marked by acute discomfort with, and reduced capacity for, close relationships as well as by cognitive or perceptual distortions and eccentricities of behavior, beginning by early adulthood and present in a variety of contexts, as indicated by five (or more) of the following:

Ideas of reference (excluding delusions of reference), odd beliefs or magical thinking that influences behavior and is inconsistent with subcultural norms (e.g., superstitiousness, belief in clairvoyance, telepathy, or "sixth sense" . . .).

In other words, if you're an eccentric introvert and believe in clairvoyance or telepathy, you might have an official psychiatric disorder. Fortunately, there are a broad array of excellent drugs that can alleviate your eccentricities and help you conform to the ideal norm where infantile fantasies like the "sixth sense" are no longer entertained. A nice course of treatment with some low potency neuroleptics (antipsychotic drugs) will help rid you of those troublesome beliefs. You may have trouble urinating and experience blurred vision for a while, but it's worth it if you can rub out those irksome beliefs in a sixth sense.

To be fair, in some forms of mental illness the ability to distinguish between reality and fantasy is so compromised that normal functioning becomes seriously degraded. A key symptom of schizophrenia is hearing voices and seeing things that no one else can hear or see. Such experiences can lead the sufferers to believe that they're extremely strong telepaths, or the world's greatest clairvoyants, or that the FBI and CIA are controlling their brains. Such beliefs can rapidly devolve into compulsions and destructive paranoia because the perceptions are uncontrolled and intrusive. Such situations are no laughing matter, and medical intervention and treatment are fully justified.

EVIDENCE

The National Science Foundation (NSF) periodically publishes a report entitled *Science and Technology Indicators,* which summarizes the state of science and technology.[3] A chapter in that report reviews the public understanding of science and technology, and one section in that chapter discusses what the NSF calls the "widespread and growing" problem of belief in pseudoscience. These are ideas or claims that superficially mimic science but do not follow standard scientific principles or rules of evidence.

A 2001 nationwide poll cited in the NSF's 2002 report asked the question, "Some people possess psychic powers or ESP. Do you strongly agree, agree, disagree, or strongly disagree?" This NSF-sponsored survey found that 60% of adult Americans agreed or strongly agreed with the statement.[4] Earlier Gallup polls taken in 1990, 1996, and 2001 showed that these percentages have been increasing over time.[5] These figures were presented in the context of demonstrating the deplorable state of science education in the United States.

This would indeed be discouraging, except that the report tiptoes around an interesting fact. When survey respondents were separated by educational level, 46% with less than a high school education agreed that some people possess ESP, but a whopping 62% with high school or more education agreed. Among the "attentive public," those defined as "very interested" in a topic, "very well informed" about it, and regularly read a daily newspaper or relevant national magazine, a healthy majority of 59% agreed. Thus, the survey actually revealed that belief in ESP was *not* explainable as a matter of poor education.

To check the NSF's findings, I examined data collected by the National Opinion Research Center, which is affiliated with the University of Chicago.[6] This Center, one of the oldest academic survey research groups in the United States, collects in its annual *General Social Survey* a wide range of questions used to form a snapshot of opinions in the United States. One of the questions asked over the years has been about psi. The specific question I was interested in asks: "How often have you felt as though you were in touch with someone when they were far away from you?" The possible answers ranged from "never in my life" to "often." I compared those answers to questions on educational achievement, which ranged from 0 to 20 years of formal education. The ignorance hypothesis predicts a negative relationship—the more education you have, the less you should believe in psi. The actual result, based on 3,880 survey responses, was not negative. In fact, it was significantly *positive,* with odds against chance of 80 to 1. This is not just the case in the United States. The same trend has been observed in Australia, France, and virtually every other country that has reported these surveys. This finding is even widely acknowledged by skeptics, who gnash their teeth about it.[7]

This is not to say that increased education has no effect on paranormal beliefs. Higher education is known to reduce belief in "religious paranormal" concepts, such as heaven, hell, the devil, and creationism. A large-scale survey in the southern United States, reported in 2003 by political scientist Tom Rice in the *Journal for the Scientific Study of Religion,*[8] compared beliefs in the religious paranormal versus psi. The survey, which involved 1,200 respondents, adopted the working hypothesis that para-

normal beliefs of both types are basically a psychological coping mechanism for people in disadvantaged social, economic, and educational conditions. Increased education was expected to correlate with decreased levels of belief for both the reli-

gious paranormal and psi. The results showed that better edu-
cated people were, as predicted, significantly less likely to believe in the religious paranormal. But contrary to the
prediction, they were significantly *more* likely to believe in psi.
This outcome was confirmed in a 2003 Harris Poll.[9]

In Sweden, a nation with one of the highest literacy rates in
the world, researchers have found that the majority of the population believes in the paranormal, and those beliefs have increased in recent decades.[10] The beliefs are not associated with
any particular organizations or social movements, and most
Swedes are not interested in institutionalized religion. Women
tend to hold more such beliefs than men, and the degree of belief is independent of educational achievement. Researcher Ulf
Sjödin, writing in the *Journal of Contemporary Religion* in 2001,
said, "It is no longer possible to neglect the paranormal as a potential ingredient in the ideologies of modern man. The survey
clearly shows that this is the case among the adult as well the
young population. So far, this has apparently been a neglected
field in the Swedish research on ideologies, and, I believe, in the
research of most other European countries."[11] Given that belief
in the paranormal in Sweden counters the ignorance hypothesis,
Sjödin pondered whether it is still reasonable "to regard such
values as deviant."

In 1999, British psychologist Chris Roe published a study in
the *British Journal of Psychology* testing the hypothesis that people
who believe in the paranormal are weak-minded or lack critical
thinking abilities.[12] Based on responses from 117 students, he
found no evidence for differences in critical thinking ability between believers and disbelievers. Other studies have confirmed
that believers have the same critical thinking skills as disbelievers.[13]

A 1997 study of paranormal beliefs by German psychologist
Uwe Wolfradt, published in the journal *Personality and Individual
Differences,* focused on the role of dissociative experiences and
anxiety in paranormal beliefs.[14] The study found that high belief

in superstitions was associated with dissociative behavior, but no such association was observed for belief in psi. High belief in psi *was* associated with characteristics like absorption, the degree to which one can shut out the rest of the world while focusing. Further analysis suggested that belief in superstition reflected a feeling of loss of control about one's life, but belief in psi was associated with the opposite, a feeling that one *is* in control. In other words, belief in psi is not due to dissociative tendencies, or to fantasy-proneness, or to the feeling of being out of control.

A study of 249 Turkish students by psychologist I. Dag, published in *Personality and Individual Differences,* confirmed that belief in psi was not a significant predictor of possible psychopathology, but belief in superstition was. Traditional religious beliefs and belief in witchcraft were related to a feeling of loss of control, but belief in psi did not support the mental deficiency hypothesis.[15]

In 2004, psychologist Anneli Goulding at Göteborg University in Sweden reported a study in *Personality and Individual Differences* based on 129 volunteers who reported strong paranormal phenomena. They filled out three questionnaires to provide a measure of schizotypy (proneness to schizophrenic behavior), mental coherence (a mental health scale), and beliefs and experiences about psi.[16] Goulding concluded that among this population of high believers, measures of schizotypy were not associated with psychological ill-health, which is contrary to the implications of the DSM-IV's definition of schizotypal personality disorder.

In summary, in spite of evidence to the contrary, some skeptics continue to assert that belief in the paranormal is best explained by ignorance or mental deficiency. Michael Shermer, publisher of *Skeptic* magazine, laments the fact that social psychology studies continue to show "no correlation between science knowledge (facts about the world) and paranormal beliefs." Shermer cited a study published in his own magazine in which the authors concluded that:

> Students that scored well on these [science knowledge] tests were no more or less skeptical of pseudoscientific claims than students that scored very poorly. Apparently, the students were not able to apply their scientific knowledge to evaluate these pseudoscientific claims. We suggest that this inability stems in part from the way that science is traditionally presented to students: Students are taught what to think but not how to think.[17]

But maybe Shermer misinterpreted this finding. An alternative view is that students are more open to their experiences than their teachers, who are defending a scientific faith that is not supported by evidence!

ON COMMON SENSE

How can beliefs so easily distort common sense? Consider something obvious like the purpose of the human heart. In the early seventeenth century, people thought that everything important to know about anatomy was already known; the Greek anatomist Claudius Galen had written it all down many centuries before. Everyone knew that the heart was a heater of the blood, and the brain, a cooler.[18] But when British physician William Harvey looked at the heart in 1628, he saw something new.[19] To him the heart looked like a pump at the center of a closed circulatory system.

Now we accept Harvey's description of the heart as common sense, and we regard Galen's earlier concept as quaintly naïve. But when Harvey's idea was first proposed, it was considered ridiculous by his medical colleagues on the European continent. They couldn't hear the heart beating as Harvey had claimed, so they saw no reason to support his proposal. A leading medical doctor of the day, Emilio Parisano of Venice, wrote the following in response to Harvey's idea:

That a pulse should arise in the breast that can be heard, when the blood is transported from the veins to the [arteries], this we certainly can't perceive and we do not believe that this will ever happen, except Harvey lends us his hearing aid. . . . He also claims that this movement produces a pulse, and, moreover, a sound: that sound, however, we deaf people cannot hear, and there is no one in Venice who can.[20]

One might think that today no one could possibly make such an obvious mistake. Unfortunately, it's not so. Beliefs can easily cause us to become blind to the obvious. Recent research on "inattentional blindness" has shown that even minor tweaks to one's expectations can cause a form of blindness. A simple experiment developed by University of Illinois psychologist Daniel Simons provides a dramatic demonstration of this effect. I've seen people take Simons's experiment and literally gasp in astonishment when they discover that they've overlooked the obvious.[21]

Simons's experiment consists of a twenty-five second video clip of six people playing a basketball game. Three are dressed in white T-shirts and three in black T-shirts. The white team is passing a basketball amongst themselves, and the black team is doing likewise. During the game, a person dressed in a black gorilla suit calmly walks into the middle of the game, beats its chest, and then walks off. The gorilla is not understated or camouflaged—it's blatantly obvious. And yet the majority of people viewing this clip do not see the gorilla provided that they're given a very simple instruction: count the number of basketballs tossed between the members wearing white T-shirts. This minor deflection of attention is sufficient to cause complete blindness to something as obvious as a gorilla. The power of deflecting attention is well known to stage magicians, who specialize in creating such illusions.

If we can so easily overlook a gorilla right in front of us, what else might we be missing? When the National Science Foundation bemoans the public's belief in topics it doesn't happen to believe in, who is being blind?

CHARACTERISTICS OF BELIEVERS

At a conference of the Institute of Noetic Sciences in 2003, we asked 465 people questions about their education, allergies, bodily sensitivities, mental practices, and unusual experiences.[22] The latter referred to experiences ranging from telepathy and precognition to reported encounters with angels and aliens. From their responses we were able to discern what kinds of people were more or less likely to report unusual experiences.

We found strong differences between men (131 respondents) and women (331). Consistent with the results of other surveys, women were less skeptical than men and reported more unusual experiences (Figure 3–1). On every type of unusual experience, from telepathy to seeing "little people," women reported higher levels of belief.[23]

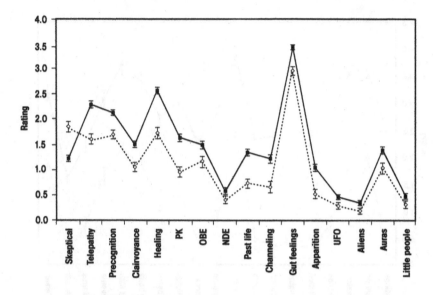

Figure 3–1. Average responses to questions about belief in unusual experiences, for women (white circles) and men (black squares), with error bars indicating the likely range of the "true" average value.

We found that left-handed and ambidextrous people were significantly more likely to believe in exceptional experiences than right-handed people, and that younger people were significantly more likely to believe than older people. Then, by comparing 55 people who reported no experiences of telepathy against 60 who frequently reported such experiences, we found a clear pattern emerging about bodily sensitivities. The "telepaths" were much more sensitive to a wide range of body and mind sensitivities (Figure 3–2). Among the no-telepathy group, half (50.9%) were female, but among the "telepaths" most (85%) were female. There were no differences in educational levels among the no-telepathy group and the telepaths.

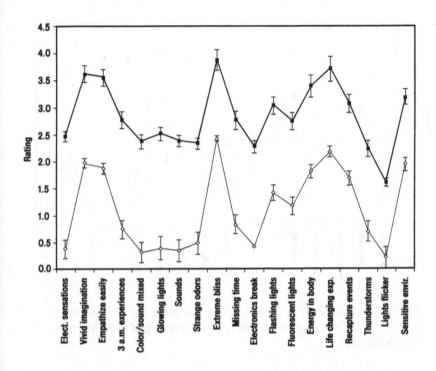

Figure 3–2. Average responses of 55 people reporting no experiences of telepathy (bottom line) and 60 people who reported it frequently (top line), with one standard error bars. The "telepaths" reported many more unusual bodily sensations, unexplained sounds and lights, sensitivities to flickering lights, thunderstorms approaching, periods of extreme bliss, and moments of missing time. The other categories in this graph are listed in this endnote.[24]

From these findings we were able to form a profile of a person very likely to report psychic experiences: a left-handed female who is thirtysomething or younger, physically highly sensitive, suffers from chronic anxiety, is somewhat introverted, makes decisions based more on feelings than logic, practices one or more of the creative arts, engages in some form of mental discipline like meditation, is open to unconventional claims, and is interested more in possibilities than in facts.

TELEPATHY OR BRAIN SEIZURES?

Some of the hypersensitivities of the "telepaths" resemble symptoms reported during temporal lobe seizures. These are sometimes referred to as "complex partial epileptic-like experiences."[25] Full-blown temporal lobe seizures can be accompanied by intense feelings of a disembodied presence, blissful religious feelings, sudden overwhelming emotions, sensory hallucinations, and numbness or electrical tickling sensations. These experiences are extremely powerful, often accompanied by overwhelming emotions, and they can lead to religious obsessions such as messianic delusions and fixations on apocalyptic ideas. But partial temporal lobe "microseizures," possibly due to exposure to strong electromagnetic fields or to inherently unstable brain structures, may lead to the very same descriptions that the "telepaths" report. This in turn would imply that the reports of telepathy are not genuine, but illusions created by misfiring brains.

Neuroscientist Michael Persinger from Laurentian University in Ontario, Canada, has for years studied the relationship between temporal lobe microseizures and reports of psychic, spiritual, and religious experiences. Using custom designed helmets studded with magnetic coils, his experiments stimulate the brain's temporal lobes with very weak magnetic fields at certain frequencies. Up to 80% of test participants wearing these helmets reportedly have experiences reminiscent of psychic and spiritual phenomena, including experiences of "vibrations, tingling sensations, odd touches, inability or reluctance to move, odd smells, odd tastes, fear or terror, intense dream-like images, and the presence of another (sentient) being."[26]

This line of research is associated with the fledgling discipline of "neurotheology," which seeks to understand the relationship of brain activity to the feelings of religious and related experiences. In its extreme form, neurotheology asserts that all psychic and spiritual experiences are illusions caused by misfir-

ing brain activity. In its more moderate and probably more accurate form, neurotheology asserts that such experiences are *correlated* with brain activity, but the causal source of these
experiences remains unknown.

A 2004 attempt at replicating Persinger's work cast doubt on
the idea that psychic and spiritual experiences can be easily
stimulated via magnetic fields. Reported under the headline,
"Electrical brainstorms busted as source of ghosts," the journal
Nature reported that psychologist Pehr Granqvist and his colleagues at Uppsala and Lund Universities in Sweden were not
able to reproduce Persinger's reports of a magnetic field effect.[27]
Under double-blind conditions, they exposed 43 students to
magnetic fields and 46 students to no fields as a control. They
found that magnetic stimulation had no effect. Over half of the
participants who reported strong spiritual experiences belonged
to the control group.

Persinger's response was that the Swedish team did not ex-
pose the participants to magnetic fields long enough to produce
an effect. British psychologist Susan Blackmore, perennial skeptic of all things paranormal and a fan of neuroscience-based explanations for unusual experiences, was also reluctant to give
up on the magnetic field theory. She reported that "When I
went to Persinger's lab and underwent his procedures I had the
most extraordinary experiences I've ever had."

While it's undoubtedly true that some forms of brain activity, especially seizures, can generate subjective feelings that
mimic psychic and spiritual experiences, it seems unlikely that
this is the sole answer. Some may regard Persinger's research as
a neuroscience approach to "explaining away" psychic experience, but Persinger himself does not accept this. For example,
Persinger's team conducted a thorough neurological investigation of the renowned psychic and artist, Ingo Swann. Swann is
the developer of a method of training remote viewing (in earlier
times this was called "traveling clairvoyance"), as used in the
U.S. government's STARGATE program of psychic spying. Swann

has repeatedly demonstrated verifiable remote viewing expertise under controlled conditions, and evidence for Swann's accurate remote viewing ability was also found in Persinger's study. So the story of psi is not as simple as a misfiring brain.[28]

LATENT INHIBITION

Maybe there's a simpler reason for the public's persistent belief in the paranormal: Maybe some of those experiences are real. And maybe the reason that creative people tend to report higher levels of belief in the paranormal is that they can see things that others can't.

An experiment supporting this idea was reported in 2003 by Harvard psychologist Shelley Carson and her colleagues in the *Journal of Personality and Social Psychology*. They examined a property known as latent inhibition. This refers to an unconscious brain process that degrades our ability to pay attention to stimuli that have had no consequences in the past.[29]

Imagine, for example, that Pavlov's dogs were exposed to ringing bells without being fed. The dogs will quickly learn to ignore ringing bells because those sounds had no meaningful consequence (i.e. no as-sociation with food). Now, Pavlov decides to train the same dogs to salivate whenever they hear a bell by ringing the bells and feeding them. Unfortunately, these dogs had already learned to ignore bells, so they're going to have a hard time unlearning the old association. Dogs that hadn't previously heard

the irrelevant bells will quickly learn to salivate. This "hard time unlearning" is due to latent inhibition.

Latent inhibition serves an important function in our brains. It allows us to perform multiple tasks, like driving a car on a busy highway while chatting with a passenger and sipping coffee without having to simultaneously pay attention to all three tasks. If we hadn't previously learned what is important to pay attention to while driving, we'd quickly become overwhelmed with information and become paralyzed with uncertainty.

Healthy people tend to have high latent inhibition. It sounds paradoxical, but the more our sensory awareness is *suppressed* by what the brain considers irrelevant, the more we remain stable and focused. If latent inhibition becomes weak it can lead to serious problems. Low latent inhibition has been studied extensively in schizophrenic patients because a key symptom of that disease is perceiving meaningful relationships everywhere, even when there aren't any. Distorted associations are associated with low latent inhibition because it reveals that the mind is having trouble ignoring irrelevant information. The 2001 Academy Award–winning movie, *A Beautiful Mind,* about the life of Nobel Laureate John Nash, suggested how this state might appear from a first-person perspective. The tag line for the movie was "He saw the world in a way no one could have imagined."

That tag line is also a good description for creative people in general, so perhaps they too exhibit low latent inhibition. Previous experiments had indeed shown that low latent inhibition is associated with the personality trait "openness to experience," which is in turn associated with divergent thinking and creativity.

Of course not all creative people are psychotic. Shelley Carson proposed that "some psychological phenomena might be pathogenic in the presence of decreased intelligence . . . but normative or even abnormally useful in the presence of increased intelligence." They tested this idea on Harvard undergraduates who were given creativity measures, IQ tests, personality tests, and a latent inhibition test. They found that the high-creativity group had significantly lower latent inhibition scores than the low-creativity group, and that the most eminently creative achievers (a subset of students who had published a novel, patented an invention, etc.) had both lower latent inhibition and higher IQ scores as compared to the other students.

Their finding supports the well-known association between genius and madness. Highly creative people have greater access to more of what the world presents, and high intelligence helps one successfully navigate through this flood of perceptions. By contrast, those with low intelligence struggle in vain, and the result may lead to psychosis. Even with high intelligence there is always the risk of becoming overwhelmed by the persistent state of expanded perception.

From this perspective, we can better understand why creative people report more psychic experiences, and why the paranormal is often associated with psychopathology. Most people who believe in psychic phenomena are not ignorant or mentally deficient. They just see farther into the depths of the world than other people do. Of course, for the sake of mental health, the goal for every creative person is learning how to peer comfortably into that abyss without becoming swallowed up by it.

CHAPTER 4
ORIGINS

If we have learned one thing from the history of invention and discovery, it is that, in the long run—and often in the short one—the most daring prophecies seem laughably conservative.

—Arthur C. Clarke

I must have been absent that day in high school when the teacher explained why history is interesting. What I recall about history classes is the odor of musty textbooks and a jumble of dates referring to politics, wars, and short periods of hysteria between wars. I found it difficult to become inspired over clashes between political and religious ideologies, or about the cavalier annihilation of nations. I *was* inspired by heroic acts of exploration and discovery, but unfortunately such stories were given short shrift in my history books.

Now that I'm a bit older, I understand why history is important. I've learned that our fundamental beliefs and assumptions, who and what we think we are, and how the engines of society, religion, and science intermesh, are all deeply rooted in the past. History is not just the story of war but a chronicle of the heroic struggle to overcome ignorance and fear. History also shows that challenges to the status quo, in whatever forms those challenges may take, always provoke violent reactions. The human body expels irritants by aggressively sneezing them out. Likewise, the body politic vigorously sneezes irritating ideas out of the scientific and academic mainstream.

Many scientists and scholars feel justified in dismissing claims of the paranormal as vestiges of an ancient, superstitious past, best left behind us. That immune response is understandable if one believes that the present scientific worldview—that collection of theories we use to explain how things work, including us—is complete. But given the lessons of history, that sort of thinking is, to put it charitably, myopic.

And this is why studying the history of psi is important. People have been reporting these phenomena for millennia and studying them for centuries. Human experiences that continue to be repeated throughout history and across all cultures, and are not due to ignorance or lack of critical thinking, demand a serious explanation.

As we examine the scientific evidence for psi in the following chapters, you may find yourself becoming lost in the trees while studying the forest. Concentrating on the details of experimental methods and statistics tends to sedate the mind into imagining that we're dealing with mere anomalies, or barely detectable statistical oddities. So it's useful to keep the larger historical picture in mind to provide a context for these experiments, and what all those minutiae add up to: evidence that common reports of extraordinary experiences are based on real, repeatable effects.

Reviewing this particular history is also useful because it's practically a secret. One of the consequences of being a scientific discipline that's not within the mainstream is that it's hardly taught anywhere. I've presented dozens of lectures on psi research at venues ranging from popular conferences to academic seminars to industrial labs and government agencies. From the questions I get after these talks, I'm occasionally dismayed to learn that some people imagine that psi research is like what we see in popular movies or in children's books like the Harry Potter series. Psi research is equated with a form of entertainment.

Others believe that psi research began with the U.S. government's formerly top-secret program of "remote viewing" or psychic spying, the existence of which was declassified in 1995. Others imagine that it started with the metal-bending Israeli psychic, Uri Geller, who first hit the media in the 1970s. Still others believe that Professor Joseph B. Rhine and his colleagues in the psychology department at Duke University invented psi research in the 1930s.

But living memory is fickle. To fully dismantle all of the mistaken beliefs and distortions would require a two thousand–page book that no one would read, so instead I'll just sketch some of the historical highlights. The complete story is a rich and fascinating tale, replete with magical rituals, occult societies, Nobel laureates, secret agents, sexual liaisons in darkened séance rooms, clandestine meetings, disinformation and false controversies, personal fears, and suicide. All the plot elements for a series of entertaining movies. And it's all true.

HISTORICAL HIGHLIGHTS IN PSI RESEARCH

In the beginning, there were no cell phones or grocery stores and life was hard. Nature was unpredictable and unforgiving. People sought ways to cope with the uncertainties of life by praying that Nature spirits would be kind to them. Magical thinking reigned supreme.

Magic has been defined as "the employment of ineffective

techniques to allay anxiety when effective ones are not available."[1] While many "old wives' tales" were futile, some were effective and based on repeated observations, presaging the origins of modern empiricism. Today we take some of those methods for granted, especially refined herbal remedies like aspirin. The use of maggots and leeches, once associated with the worst horrors of medieval medicine, is also back in vogue because those ancient folk remedies can do things that the medical miracles of today still cannot surpass. The rising interest in alternative medicine, especially botanical herbs, acupuncture, and perhaps homeopathy suggests that in the modern rush to adopt synthetic miracle drugs some effective traditional remedies may have been prematurely dismissed as superstitions. Some of those old wives were probably smarter than we know.

As magical concepts evolved, they tended to fall into two basic classes: natural and supernatural. The former pertained to properties inherent in objects themselves, the latter to acts of superior, invisible beings. The study of natural magic presaged science, and the concept of supernatural magic was largely subsumed within religious doctrine.

Millennia passed. Knowledge about the natural world advanced slowly. In 2000 BC, Egyptians practiced dream incubation as a technique for evoking oracles. They slept in special temples in the hopes of inducing divinely inspired dreams.[2] A few hundred years later in China, oracles would toss tortoise shells into the fire and read the resulting cracks in the shells as omens about future events. The predictions and outcomes of these divinations were inscribed on the shells.[3] The 50,000 known Shang dynasty "Oracle Bones" are among the earliest known psi experiments and earliest forms of written language. They indicate not only that oracles were commonplace, but also that oracles could—and should—be put to the test by comparing predictions with outcomes.

In 650 BC, one of the longest-lived businesses in history began. The Delphic oracle at the Temple of Apollo in Greece

lasted for seven hundred years. The god Apollo was said to foretell the future through his priestess, the Pythia. She inhaled vapors rising up through cracks in the Temple's floor to induce an altered state of consciousness, and then she responded to the questions of visitors while in a trance. An official interpreter inscribed her resulting moans and mumblings.[4] It's difficult to know how effective the Delphic oracles were in forecasting the future, as few written records remain. Fortunately, Herodotus did carefully document one test case. He wrote that King Croesus of Lydia wished to consult an oracle, and he knew that most of the oracles of the day were fakes. So the king devised a test to find one with genuine skill. The Pythia of the Temple of Apollo was the only oracle who responded to his experiment with the correct answer. She said through her interpreters, in traditional hexameter verse:

I can count the sands, and I can measure the ocean;
I have ears for the silent, and know what the dumb man meaneth;
Lo! on my sense there striketh the smell of a shell-covered tortoise,
Boiling now on a fire, with the flesh of a lamb, in a cauldron—
Brass is the vessel below, and brass the cover above it.[5]

In fact, King Croesus had taken a tortoise and a lamb, cut them into pieces, and boiled them together in a brazen cauldron covered with a lid, also made of brass. On the basis of the Delphic oracle's accuracy, Croesus consulted her about what would happen if his army invaded Persia. She replied that if he did this, it would "destroy a great empire." Croesus assumed this meant his invasion would crush Persia, but unfortunately he didn't verify that flattering interpretation. As history shows, he did indeed destroy a great empire—except it was his own.[6] When dealing with oracles, it's a good idea to check your assumptions.

In Greece, Democritus believed in dream telepathy and divination,[7] but Aristotle was less certain. He wrote:

> As to divination which takes place in sleep, and is said to be based on dreams, we cannot lightly either dismiss it with contempt or give it implicit confidence. The fact that all persons, or many, suppose dreams to possess a special significance, tends to inspire us with belief, based on the testimony of experience . . . . Yet the fact of our seeing no probable cause to account for such divination tends to inspire us with distrust.[8]

Cicero agreed. In a typically caustic remark, his comment about Democritus was, "I never knew anyone who talked nonsense with greater authority."[9]

Half a millennium later, in 1484, Pope Innocent VIII published a bull against witches, followed by the notorious document, *Malleus Maleficarum* (The Witch Hammer). This made witchcraft a capital crime and inspired a madness called the witch hunt, which became a wildly popular sport throughout Europe. A hundred and twenty years later, King James I of England issued the Witchcraft Act: "An acte against conjuration witchcrafte and dealinge with evill and wicked Spirits." Practicing witchcraft was now against the law as well as against church doctrine.

Two decades later, in 1627, Sir Francis Bacon published *Sylva Sylvarum: or A Naturall Historie In Ten Centuries*. Bacon was an au-

thor, barrister, and eventually Lord Chancellor of England. He is credited with developing the basis of empirical reasoning, one of the core concepts underlying the power of the scientific method. Before Bacon, anyone looking for a reliable answer to a question about Nature would have been advised to consult Aristotle, as Aristotle's authority had not been questioned for over two thousand years. In *Sylva Sylvarum,* Bacon proposed that mental intention (his actual phrase was the "force of imagination") could be studied on objects that "have the lightest and easiest motions. And therefore above all, upon the spirits of men," by which he meant the emotions. Bacon continued, "As for inanimate things, it is true that the motions of shuffling of cards, or casting of dice, are very light motions," presaging the use of cards and dice and other random physical systems in psi experiments.

Bacon further proposed that in studies on the "binding of thoughts," or what we'd now call telepathy, that "you are to note whether it hit for the most part though not always," anticipating the use of statistical techniques. Further, he noted that one might be more likely to succeed in such tests "if you . . . name one of twenty men, than . . . one of twenty cards," that is, tasks involving meaningful targets might be more effective than tasks involving the guessing of simple playing cards. Bacon's ideas were not only 300 years ahead of their time. They show that interest in testing psi was among the very first proposed uses of science.

Half a century after *Sylva Sylvarum* was published, the infamous witch trials began in Salem, Massachusetts. That hysteria resulted in the deaths of 19 innocent people and the accusation of hundreds more.[10] A decade later, the Massachusetts General Court officially declared the trials unlawful.

Time passed. The scientific revolution in Europe was accelerating. Ideas promoted by Bacon and other luminaries such as Copernicus, Kepler, Galileo, Descartes, and Newton began to take hold, and the fledgling science proliferated.

EIGHTEENTH CENTURY

Emanuel Swedenborg was a renowned metallurgist and mystic in the mid eighteenth-century. Among his many scientific accomplishments, Swedenborg displayed an astonishingly modern understanding of brain functioning. Two hundred years before the neurosciences became a scientific discipline, Swedenborg correctly described sensation, movement, and cognition as functions of the cerebral cortex, the function of the corpus callosum, the motor cortex, the neural pathways of each sense organ to the cortex, the functions of the frontal lobe and the corpus striatum, circulation of the cerebrospinal fluid, and interactions of the pituitary gland between the brain and the blood.[11] On the afternoon of June 19, 1759, he arrived in Göteborg, Sweden. At a dinner party that evening, he suddenly announced to his friends that he was having a vision of Stockholm burning, about 300 miles away. Later that evening he told them that the fire stopped three doors from his home. The next day, the mayor of Göteborg, who heard about Swedenborg's surprising pronouncement, discussed it with him. The following day, a messenger from Stockholm arrived and confirmed that Swedenborg's vision was correct.[12]

A few decades later, the American colonies declared their independence from Great Britain. While George Washington was battling the British, Austrian physician Franz Anton Mesmer was advancing the concept of "animal magnetism." At the time, electricity and magnetism were evoking great interest as newly discovered, still-mysterious forces of nature. Mesmer proposed that animal magnetism was a biological force analogous to those physical forces.[13] Mesmer's ideas helped to develop the origins of hypnosis, psychoanalysis, and psychosomatic medicine.

French aristocrat Armand Marie Jacques de Chastenet, known as the Marquis de Puységur, was one of Mesmer's early students. Puységur accidentally discovered the first method known to reliably evoke psi phenomena. He called his discovery "magnetic somnambulism," a type of "sleepwalking" state we now call deep hypnosis. He found that some somnambulists showed the full range of paranormal skills, including telepathy, clairvoyance, and precognition.

The explosion of popular interest in Mesmer and Puységur's methods outraged the physicians of the day. Their indignation triggered an investigation by the French Academy of Sciences in 1784, chaired by Benjamin Franklin, who had been sent to France by the American Congress in 1776 in hopes of gaining France's support for the American Revolution. The French Academy was charged with evaluating the scientific status of mesmerism. A month later, a second commission was formed under the auspices of the French Royal Society of Medicine. It was asked to determine whether mesmerism was useful in treating illness, regardless of whether there was any scientific explanation for it. After numerous tests, both commissions concluded that there wasn't any evidence for the "magnetic fluid" proposed by Mesmer, and that all of the observed effects could be attributable to the imagination (the placebo effect). However, the Royal Society's conclusion wasn't unanimous; a minority report declared that some healing effects could not be attributed solely to imagination.[14]

NINETEENTH CENTURY

A half-century later, mesmerism was still raging throughout Europe, so the French Royal Society of Medicine launched a new investigation. This time the report was uniformly favorable not only to mesmerism but also to the somnambulistic psi phenomena reported by Puységur. The report ended with a recommendation that the Royal Society continue to investigate these phenomena. For the next five years studies took place and the commissioners described many examples of psi phenomena that they had personally witnessed.[15] This was one of the first major government-sponsored scientific investigations of psi effects that had an entirely positive outcome. And it wasn't just the Royal Society that was impressed. Jean Eugene Robert-Houdin, the most famous stage magician of his day (from whom Ehrich Weiss, better known as "Houdini," would later adopt his stage name), "confessed that he was completely baffled" about a somnambulist named Alexis, who displayed the clairvoyant ability to read playing cards while blindfolded.[16]

In the United States, in 1848, as war with Mexico was winding down and conflict between the Northern and Southern states was heating up, two young sisters named Margaretta and Catherine Fox, of Hydesville, New York, reported that they had established communications with spirits who were apparently responding to their questions with rapping sounds. Similar poltergeist ("noisy ghosts") outbreaks had been reported from antiquity, but this one caught the public fancy and the spiritualism craze quickly spread throughout the United States and Europe. Séances to contact the dead became a wildly popular parlor game. Con artists took advantage of the public interest by offering performances staged as legitimate séances. Many of these so-called mediums were ultimately unmasked as frauds, but some were genuine enigmas. A Scottish medium named Daniel Dunglas Home astounded European audiences by levitating in plain view, with many witnesses present. He performed this and other feats not matched before—or since.[17] Despite

dozens of performances, Home was never caught cheating. His abilities remain a mystery.

Distinguished British scientist Sir William Crookes, who was President of the Chemical Society, the Institution of Electrical Engineers, the British Association for the Advancement of Science, and vice president of the (British) Royal Society was so intrigued by Home's performances that he created special laboratory equipment to study him. Crookes was impressed with the results and considered Home to have genuine abilities. Sir Francis Galton, cousin of Charles Darwin, polymath scientist and inveterate skeptic, offered his opinion of Crookes's investigations of physical mediumship:

> Crookes, I am sure, so far as is just for me to give an opinion, is thoroughly scientific in his procedure. I am convinced that the affair is no matter of vulgar legerdemain and believe it is well worth going into, on the understanding that a first rate medium (and I hear there are only three such) puts himself at your disposal.[18]

In 1850, California became the thirty-first state in the United States, and Nathaniel Hawthorne published *The Scarlet Letter.* On October 22 of that year, German physicist Gustav Theodor

Fechner had an inspiration that led to the origins of modern experimental psychology and psychophysiology. Fechner's insight was based on his belief that mind and matter arise from the same, nonmaterial, spiritual source. In his attempt to refute materialism by demonstrating relationships between mind and matter, he placed the new discipline of psychology on firm scientific grounds. Despite his many scientific achievements, his less celebrated colleagues considered his mystical inspirations the eccentricities of a mad genius.[19]

A quarter century later, in 1876, the American Civil War had come and gone, the Heinz company began selling ketchup, and Lt. Colonel George Custer and 647 men of the Seventh Cavalry were defeated by Cheyenne and Sioux Indians at the Battle of the Little Bighorn. That same year in England, physicist Sir William Barrett from the Royal College of Science in Dublin, Ireland, presented his research on "thought transference" to the British Association for the Advancement of Science.[20] Six years later, Barrett helped found the London-based Society for Psychical Research (SPR), the first scientific organization in the world established for the study of psi phenomena. In his inaugural report to the SPR's Committee on Thought-Reading, Barrett complained about the prejudice against these topics within the scientific community:

> The present state of scientific opinion throughout the world is not only hostile to any belief in the possibility of transmitting a single mental concept, except through the ordinary channels of sensation, but, generally speaking, it is hostile even to any inquiry upon the matter. Every leading physiologist and psychologist down to the present time has relegated what, for want of a better term, has been called "thought-reading" to the limbo of exploded fallacies.

Fortunately, not everyone was blinded by prejudice. Many prominent members of British, European, and American science, scholarship, and politics became members of the SPR. The roster included physicist Sir Oliver Lodge, best known for his contributions to the development of wireless telegraphy, and physicist Baron Rayleigh, who was married to Evelyn Balfour, sister of Arthur James Balfour, the prime minister of Britain. Rayleigh was later awarded the Nobel Prize for his discovery of the inert gas argon. American members of the SPR included astronomer Samuel P. Langley, Director of the Smithsonian Institution; psychologist William James of Harvard University; astronomer Simon Newcomb, President of the American Association for the Advancement of Science; and Edward C. Pickering, Director of the Harvard Observatory.[21]

A few years after the formation of the SPR, French physiologist Charles Richet published an article describing his experiments on telepathy using playing cards. He introduced "a method which is extremely rarely in usage in the sciences, the method of probabilities."[22] This was the first paper to use statistical inference for studying telepathy in the general population. Richet concluded that there did exist, "In certain persons at certain times, a faculty of cognition which has no relation to our normal means of knowledge."[23] Richet would later win the Nobel Prize for his research on anaphylaxis, and at one point he served as president of the SPR.

In light of Richet's claims about telepathy, the eminent British economist F. Y. Edgeworth was asked by members of the SPR to provide his opinion of Richet's use of statistical inference. Edgeworth published two papers in the *Proceedings of the SPR*, which have been 'described as "fine papers, beautiful enough almost to justify the entire subject of parapsychology."[24] Edgeworth, a staunch skeptic, confirmed that Richet's card-guessing experiments were not due to chance, as they resulted in odds against chance of 25,000 to 1. He concluded that Richet's claims:

May fairly be regarded as physical certainty, but is silent as
to the nature of that agency—whether it is more likely to be
vulgar illusion or extraordinary law. That is a question to be
decided, not by formulae and figures, but by general philosophy and common sense.[25]

About a decade later, in 1886, the Apache chief Geronimo,
who had achieved legendary status by surviving 15 years of battle with the Mexican Army and U.S. Cavalry, finally surren-
dered. Coca-Cola, so-named because it contained genuine
cocaine, was introduced as "a valuable brain-tonic and cure for
all nervous afflictions."[26] In Germany, physicist Heinrich Hertz
noticed that certain spark coils he was experimenting with
worked more easily if ultraviolet light was shining on them. He
didn't know it yet, but this was the first observation of the photoelectric effect, a phenomenon that helped launch modern
physics and the theory of quantum mechanics. Hertz was a
long-term corresponding member of the Society for Psychical
Research.

A decade later, British physicist J. J. (Joseph John) Thomson
discovered the electron, for which he was awarded the Nobel
Prize in 1906. Two years later, in an address to the British Association for the Advancement of Science, Thomson speculated
that electromagnetic fields were carriers of information between
people, and hence they provided a physical mechanism for
telepathy. Sir J. J. Thomson served as a member of the Govern-
ing Council of the SPR for 34 years.

As the nineteenth century drew to a close, German psychiatrist Sigmund Freud wrote his first paper on the occult. His second paper was published in 1904 and a third in 1919. His initial
attitude was entirely negative, associating the occult solely with
superstition. Later his attitude changed to caution and intellectual curiosity. By 1921 he wrote, "It no longer seems possible to
brush aside the study of so-called occult facts."[27]

TWENTIETH CENTURY

At the dawn of the twentieth century, German physicist Max Planck postulated that energy was radiated in tiny, discrete units, which he called quanta. The quantum era was born. Two years later, Nabisco's *Animal Crackers* first went on sale, and German psychiatrist Carl Jung wrote his doctoral thesis on a psychological study of a medium. Later, in an instance of what Jung called "synchronicity" in a therapeutic setting, Jung described the case of a young patient whose extreme rationalism was blocking psychoanalytic treatment. As the story goes,

> One day [Jung] sat opposite her with his back towards the window, listening to the flow of her rhetoric. The night before she had told him a dream about somebody who had given her a costly golden scarab as a present. At that very moment Jung heard something gently tapping on the window. It was a big insect trying to get into the dark room. He let it in, caught it, and it turned out to be a common rose chafer—a beetle closely resembling a golden scarab. He handed it to the patient: "Here is your golden scarab." "This experience," Jung notes, "punctured the desired hole in her rationalism and broke the ice of her intellectual resistance."[28]

The following year, Frederic Myers of the SPR published one of the first scholarly volumes investigating the possible survival of consciousness, entitled *Human Personality and Its Survival of Bodily Death*. At about the same time, Marie and Pierre Curie isolated radium, for which they would later receive the Nobel Prize. The Curies also started to attend séances by famed Italian medium, Eusapia Palladino.

In 1905, the picture postcard and the Popsicle were born, and Apache Chief Geronimo rode as an invited guest in President Theodore Roosevelt's inaugural parade in Washington, DC. In Switzerland, an unknown twenty-six-year-old patent clerk named Albert Einstein published three papers that would change the face of physics for the next century. One of those papers presented an explanation for the photoelectric effect discovered earlier by Heinrich Hertz. Einstein received the 1921 Nobel Prize for his contribution.

In 1911, the British Empire covered twenty percent of the world's land area. Thomas Welton Stanford, the brother of the founder of Stanford University, donated £20,000 to Stanford "to a fund which shall be known as the 'Psychic Fund.'" This was to be used "exclusively and wholly for the investigation and advancement of the knowledge of psychic phenomena and the occult sciences . . ." When Thomas Stanford died 20 years later, his will left an additional $526,000 (about $10 million in 2005 U.S. dollars) to this fund. The first Thomas Welton Stanford Psychical Research Fellow was a man named John Edgar Coover; he held the chair from 1912 to 1937. After Coover, no one ever again permanently held this research fellowship (which still exists).

Most of Coover's research was published in one book, in 1917.[29] He claimed not to find any evidence in support of psi, but only because he set his threshold for positive evidence at the stupendously high odds of 50,000 to 1. He dismissed any evidence that didn't surpass that level. His book included a description of the first example of a randomized, blinded, controlled study, a technique that has since become the gold standard in the psychological and medical sciences. For that reason we can forgive him for ignoring the terms of his appointment for the remaining 20 years he spent at Stanford. His lack of interest in pursuing psi research is all the more puzzling given that we now know that his claim of finding no evidence was questionable. Coover's book shows that he did obtain significantly positive evidence for telepathy in his experiments, with respectable odds against chance of 167 to 1.

Just as Coover was beginning the Stanford Fellowship, Sir J.J. Thomson at Cambridge University hired Francis Aston as an assistant. Aston had read a 1908 book, *Occult Chemistry,* by British Theosophists Annie Besant and Charles Leadbeater. In that book, Besant and Leadbeater described their clairvoyant vision of the internal structure of atoms, including a new form of the element neon, which they called meta-neon. They claimed that meta-neon had an atomic weight of 22.33. In 1912, Aston discovered a substance at that atomic weight while analyzing neon gas. He also dubbed it meta-neon in a paper presented to the annual meeting of the British Association for the Advancement of Science. Aston's discovery was later labeled an "isotope," and it became a key discovery about atomic structure (which years later, led to the development of the atomic bomb). Aston received the 1922 Nobel Prize for his work, but in his acceptance speech he somehow forgot to mention the inspiration for his discovery.[30]

Around this time, much of the Northern Hemisphere was plunging headlong into the First World War. In 1917, the United States Selective Service Act created a draft in preparation for military action. In the midst of all the excitement, psychologist Leonard Troland at Harvard University obtained successful results with one of the first automated ESP testing machines.[31] A few years later, as the War was winding down, the Institut Metapsychique International was founded in Paris, France. Its first President was Nobel Laureate physiologist Charles Richet. A few years later, French researcher René Warcollier described some of the first highly successful picture-drawing psi experiments in a book entitled *La Télépathie.*

Between the euphoria following the end of the First World War and the despair of the U.S. stock market crash of 1929, the distinguished British statistician Sir R. A. (Ronald Aylmer) Fisher solved problems in statistical inference for use in card guessing tests by psi researchers,[32] and German physicists Werner Heisenberg, Max Born, and Pascual Jordan developed matrix mechanics, the first version of the mathematics of quantum mechanics. Jordan, like his Nobel Prize–winning colleague

Wolfgang Pauli, was vitally interested in psi phenomena. Jordan would later write in the *Journal of Parapsychology*,

> The existence of psi phenomena, often reported by former authors, has been established with all the exactness of modern science by Dr. Rhine and his collaborators, and nobody can any longer deny the necessity for taking the problem seriously and discussing it thoroughly in relation to its connections with other known facts.[33]

Sigmund Freud had also become increasingly interested in psi. In writing to a friend, he explained why his earlier public stance on telepathy had been so reserved, and why he had changed his mind:

> As you remember I already expressed a favorable bias toward telepathy during our trip to the Harz. But there was no need to do so publicly; my conviction was not very strong, and the diplomatic aspect of preventing psychoanalysis from drawing too close to occultism very easily retained the upper hand . . . . In the meantime, however, my personal experience through tests, which I undertook with Ferenczi and my daughter, have attained such convincing power over me that diplomatic considerations had to be relinquished.[34]

Just before the stock market crash of 1929, biologist Joseph Banks Rhine started a psi research program at Duke University, sponsored by the chair of the psychology department, William McDougall, who had founded the *British Journal of Psychology*. Rhine's parapsychology research continued at Duke until 1965. Just before Rhine left Duke, in 1962, which incidentally was the same year that instant mashed potatoes was introduced, Rhine established the Foundation for Research on the Nature of Man (FRNM) with the assistance of benefactor Chester Carlson, founder of the Xerox Corporation. The FRNM ran for 40 years, from 1962 to 2002, when it was renamed the Rhine Research Center.

During the short respite between the stock market collapse and the start of the Second World War, social activist and author Upton Sinclair published *Mental Radio*. This popular book described Sinclair's successful picture-drawing telepathy tests with his wife, Mary Craig Sinclair. A few years later, Rhine's book, *Extra-Sensory Perception* was published, evoking great interest among academic circles and the general public. Soon after that, British psychologist G. N. M. Tyrrell reported the development of an ESP testing machine with features that would later become essential design principles in psi experiments, including random target selection and automatic data recording.[35] And then in 1937, the *Journal of Parapsychology* began publication, founded by J. B. Rhine and his colleagues.

Just before the Second World War erupted, Sir Hubert Wilkins and Harold Sherman conducted a remarkable long-distance experiment in clairvoyance.[36] Wilkins was an Australian photographer and naturalist who gained fame for exploring the world in airplanes and submarines. Sherman was a popular author and playwright with a long-term interest in psychic phenomena. The experiment was sparked by the loss of a Russian plane somewhere in the Arctic off the coast of Canada. Given Wilkins's knowledge of the Arctic and his piloting skills, he was asked by the Russian government to see if he could find the missing plane. He agreed, and Wilkins and Sherman decided to use this opportunity to see if Sherman could "tune in" to Wilkins at a distance. On a daily basis, Sherman used clairvoyance to "see" what was happening to Wilkins and his team. Wilkins, in turn, kept a daily log of each day's events, which was later compared against Sherman's perceptions.

Intercontinental communication was sporadic at best in 1938, and communication with Wilkins, who was usually flying a small plane off the coast of Alaska, was impossible.

Weeks would often pass from the time when Wilkins wrote his daily reports to when they were received in New York City. To ensure that the experiment was conducted fairly, each day Sherman deposited copies of his nightly impressions to third-party witnesses, all of whom later attested that the recordings were in their hands before Wilkins's log was received.

As an example of the similarities in their reports, on November 30, 1938, Wilkins and his team were in Aklavik, in the Canadian Northwest Territories. This was the middle of nowhere, in the middle of the winter, above the Arctic Circle. Within the small settlement of Aklavik, at one point Wilkins and his men were invited to attend a party at the local hospital. They did so, and later that evening two of his crew went to the basement where they were surprised to find Ping-Pong tables. They played Ping-Pong with some nurses and had a grand time.

That evening in New York City, some 3,000 miles away, Sherman recorded his nightly clairvoyant vision as follows: "I

received a strong impression of 'Ping-Pong balls,' for some reason, and found myself writing: "sudden flash of Ping-Pong—is there table in town where people play? Can't account [for] this unusual impression . . . ." Wilkins later noted, after reading Sherman's impressions of this day, that "[Sherman] would have hardly guessed that we would be playing Ping-Pong in the Arctic." Dozens of such correspondences are described in their book.

As the Second World War raged in Europe, J. B. Rhine and his colleagues from Duke University published the book *Extrasensory Perception After Sixty Years*.[37] It analyzed in detail all known ESP card-guessing experiments conducted over the sixty years from 1880 through 1939. Meanwhile, during the war in Europe, British psychologist Whately Carington conducted picture-drawing experiments with large groups of people. His goal was to develop a repeatable free-response clairvoyance experiment.[38] The results were highly successful.

After the war, in 1949, the Soviet Union tested its first atomic bomb, and Rodgers and Hammerstein's musical *South Pacific* opened on Broadway. Albert Einstein compared quantum theory's prediction about entangled particles to telepathy.[39] He used this analogy to imply that quantum theory must be incomplete because he couldn't believe that any separated objects could be entangled, either at the atomic or human scales. Einstein was astonishingly correct about many things. This wasn't one of them.

In 1950, the first fully automatic telephone answering machine was invented at Bell Laboratories, and the first credit card, the Diners Club, was launched. In Freiburg, Germany, the Institut für Grenzgebiete der Psychologie und Psychohygiene (Institute for Border Areas of Psychology and Mental Health) was founded by psychologist and physician Hans Bender. It would become a key psi research organization in Europe.

At the same time in England, mathematician Alan Turing, a

seminal figure in the foundations of modern computer science and the mastermind who helped break the German Enigma cryptograph machine during the Second World War, wrote about the evidence for psi:

> I assume that the reader is familiar with the idea of extrasensory perception, and the meaning of the four items of it, viz., telepathy, clairvoyance, precognition and psychokinesis. These disturbing phenomena seem to deny all our usual scientific ideas. How we should like to discredit them! Unfortunately the statistical evidence, at least for telepathy, is overwhelming. It is very difficult to rearrange one's ideas so as to fit these new facts in. Once one has accepted them it does not seem a very big step to believe in ghosts and bogies. The idea that our bodies move simply according to the known laws of physics, together with some others not yet discovered but somewhat similar, would be one of the first to go. . . . Many scientific theories seem to remain workable in practice, in spite of clashing with ESP; that in fact one can get along very nicely if one forgets about it. This is rather cold comfort, and one fears that *thinking* is just the kind of phenomenon where ESP may be especially relevant.[40]

In 1951, physicist Edward Teller was preparing to test the hydrogen bomb, and there were political revolutions in Thailand, Panama, Bolivia, and Argentina. In England, the Witchcraft Law of 1735 was finally repealed and replaced by the Fraudulent Mediums Act. In the United States, Eileen Garrett, a talented medium who had worked extensively with scientists, established the Parapsychological Foundation in New York City.

In 1953, Dow Chemical introduced the ever-popular Saran Wrap, and Sir John Eccles introduced psilike, mind-matter interaction effects in his model of mind-brain interaction.[41] A decade later Eccles was awarded the Nobel Prize. In Holland, psychologist W. H. C. Tenhaeff established the Institute for Parapsychology in Utrecht, which was affiliated with the University of Utrecht. J. B. Rhine at Duke University received a grant from the United States Office of Naval Research to investigate ESP in animals.

In 1957, the year that General Foods introduced *Tang*, an orange-flavored instant breakfast beverage, the Commonwealth of Massachusetts belatedly apologized for the Salem witch trials of 1692. The Parapsychological Association, an international organization of scientists and scholars, was founded, and Czech physician Štěpán Figar measured fingertip bloodflow in pairs of isolated people to test for unconscious telepathic connections. In Figar's test, neither person was aware of the other, nor were they told the purpose of the experiment. He found that when one of the pair was asked to perform mental arithmetic, the blood pressure of the other person changed noticeably.[42] This was one of the first experiments investigating unconscious forms of telepathy between isolated people.

In 1963, the one billionth McDonald's hamburger was served by founder Ray Kroc on Art Linkletter's popular television show,[43] U.S. President John F. Kennedy was assassinated in Dallas, Texas; and Russian physiologist Leonid Vasiliev published the book *Experiments in Mental Suggestion*. Vasiliev had pioneered Russia's exploration of "remote hypnosis" in the 1920s and 1930s, replicating the somnambulistic phenomena discovered more than a century before by the Marquis de Puységur. Vasiliev demonstrated that somnambulists could be induced to fall into deep trance states when given hypnotic suggestions from a distance, in some cases thousands of miles away. This book was important not only because of the phenomena described, but because Vasiliev described how his experiments were taken seriously at the highest levels of the Russian government and scientific establishments.

In 1964, Martin Luther King Jr. was awarded the Nobel

Peace Prize, the Beatles song "I Want to Hold Your Hand" became the #1 pop song in the United States, and psychologist Montague Ullman launched a series of dream telepathy studies at the Maimonides Medical Center in Brooklyn, New York. That same year in Europe, Irish physicist John Bell mathematically proved that quantum theory requires "spooky action at a distance." This famous proof would become known as Bell's theorem, and some physicists regard it as the most profound scientific discovery of the twentieth century.

The following year, the journal *Science* published its usual array of scientific papers, including a distinctly atypical article entitled "Extrasensory electroencephalographic induction between identical twins." Two researchers from the Department of Ophthalmology at Jefferson Medical College in Philadelphia reported striking—one might say spooky—correspondences in the electroencephaolographs (EEG) of distance-separated pairs of identical twins. The notion of entangled minds was born.

In 1969, the Beatles' album *Abbey Road* was released; it would become one of the best-selling albums of all time. Apollo 11 astronaut Neil Armstrong became the first man to step on the moon, and Helmut Schmidt, a German-American physicist at Boeing Scientific Laboratories, published a paper about an automated psychokinesis experiment using an electronic "coin-flipper" circuit called a random number generator (RNG). This would become a model for one of the most frequently replicated psi experiments over the next several decades. The Parapsychological Association was elected an official affiliate of the American Association for the Advancement of Science in 1969, marking the first mainstream acknowledgement of psi research as a legitimate scientific enterprise.

In 1972, the first successful video game (Pong) was released, Nike running shoes were first sold, and Richard Nixon became the first U.S. president to visit the Soviet Union. Physicists

Harold Puthoff, Russell Targ, and Edwin May began a program of classified research on psi phenomena for numerous U.S. government agencies, and physicists Stuart Freedman and John F. Clauser published a successful experimental test of Bell's theorem. The following year, Apollo 14 astronaut Captain Edgar Mitchell, the sixth man to walk on the moon, founded the Institute of Noetic Sciences. Mitchell had conducted a successful ESP card experiment from the Apollo 14 space capsule.[44]

In 1979, the Sony Walkman was introduced, there was a meltdown at the Three Mile Island nuclear power plant near Harrisburg, Pennsylvania, and Charles Honorton founded the Psychophysical Research Laboratories in Princeton, New Jersey, with the support of James McDonnell of McDonnell-Douglas aircraft. The Dean of the School of Engineering and Applied Science at Princeton University, Robert Jahn, established another psi research laboratory in Princeton. The Princeton Engineering Anomalies Research (PEAR) program would become of one the world's principal psi research groups. A few years later, Jahn published the results of his laboratory's initial experiments in the journal *Proceedings of the Institute of Electronic and Electrical Engineers,* and French physicist Alain Aspect and his colleagues at the Institut d'Optique in Orsay, France, published the first widely accepted evidence that spooky action at a distance exists. The idea of quantum entanglement was no longer a theoretical possibility, but an experimental fact.

In 1981, members of the U.S. Congress asked the Congressional Research Service to assess the scientific evidence for psi. The review was prompted by concerns that if psi effects were genuine, we'd have to assume that foreign governments would exploit it. Over the next 15 years, the U.S. Army Research Institute, the National Research Council, the Office of Technology Assessment, and the American Institutes for Research prepared similar reports (the latter at the request of the Central Intelligence Agency). While quibbling

over fine points of interpretation, all five of these reviews concluded that some of the experimental evidence for psi warranted serious study.

In 1985, crack cocaine was becoming a major epidemic,
Coca-Cola's soft-drink campaign for "New Coke" was a spectacular failure, and Yale University psychologist Irvin Child
published a positive article on "ESP in Dreams" in the journal
American Psychologist.[45] In England, author Arthur Koestler and
his wife Cynthia had bequeathed funds to establish an endowed
Chair of Parapsychology at a British university. The Chair was
adopted by the University of Edinburgh, and the first holder of
the Chair was American psychologist Robert Morris, who held
the post from December 1985 until his death in August 2004.
Besides being a major figure in parapsychology, Morris served
as president of the Psychology Section of the British Association
for the Advancement of Science. By 2004, the nearly 50 students who had earned their PhDs under his aegis had helped to
establish psi research as a legitimate topic of academic study in
the United Kingdom.

In 1987, one of the longest and most expensive trials in U.S.
history began. The McMartin Preschool abuse trial involved a
group of preschool teachers who were falsely accused of abus-
ing children in Satanic rituals. Like the Salem witch trials and
the communist witch-hunts of the McCarthy era in the United
States in the early 1950s, this case was a contemporary re-
minder that moral hysteria associated with fears of the supernatural (and its political equivalents) is always uncomfortably
close to the surface of the human psyche.[46] That same year, psychologist Ramakrishna Rao, director of the Foundation for Research on the Nature of Man, along with psychologist John
Palmer, published an extensive, positive article on psi research
in the journal *Behavioral and Brain Science.*

In 1989, the Cold War thawed, reports of "cold fusion"
were greeted with wild acclaim by the press and then immediately attacked by the scientific mainstream, Teenage Mutant

Ninja Turtles were a smash hit, and Sony Labs researcher Yoichiro Sako approached one of Sony's two founding fathers, Masaru Ibuka, about establishing a psi lab within Sony. Ibuka agreed, and the "ESPER" lab began operations. Some years later Ibuka died, and the lab closed. When asked about the ESPER lab's research, Sony Labs spokesman Masanobu Sakaguchi reportedly said: "We found out experimentally that yes, ESP exists, but that any practical application of this knowledge is not likely in the foreseeable future."[47]

In 1994, interest in the World Wide Web began to explode and the bizarre spectacle of O. J. Simpson's "slow speed" car chase on the Los Angeles freeways was broadcast live around the world. The Bial Foundation, part of a pharmaceutical company in Portugal, began offering research grants in parapsychology. This Foundation would soon become one of the world's major independent sponsors of psi research. The following year, the U.S. government's secret program of psi research and applications, code-named STARGATE at the time, was made public by the CIA. At the same time, a new psi research program supported by the Japanese Government, and headed by Mikio Yamamoto, began operations at the National Institute of Radiological Sciences, part of Japan's Science and Technology Agency. That program closed in 2005, when Yamamoto retired.

In 1998, as the U.S. economy began to expand at an unprecedented rate due to the rise of the Internet, the first U.S. patent for a psi-operated electronic switch was granted (number 5830064). The patent was based on the mind-matter interaction research of the Princeton Engineering Anomalies Research Laboratory, which we'll discuss later. That same year I started a psi research program at Interval Research Corporation, a consumer electronics research lab in Silicon Valley.[48] Two years later, University of Amsterdam psychologist Dick Bierman began another psi research program at Star-

Lab, an industrial research lab in Belgium.[49] These fledgling technology-oriented developments foreshadowed a change in how psi is perceived among leading-edge thinkers. No longer is it viewed as unthinkable, or as a meaningless anomaly. In-

stead, psi is being regarded as a genuine, albeit poorly understood human facility that, if we can figure out ways of using it reliably, will undoubtedly become the next trillion-dollar business. This pragmatic shift is beginning to trump outdated skepticism.

There are always those who will vigorously deny the possibility of new-fangled inventions, but as those naysayers enjoy refreshing beverages while watching television in their comfortable homes, they might consider that if creative scientists and engineers weren't constantly imagining impossible futures, then we'd all still be living in damp caves and eating grubs for dinner.

Science fiction author H. G. Wells once said that history is a race between education and catastrophe. Philosopher George Santayana offered similar advice, warning that those who cannot remember the past are condemned to repeat it. There's a similar imperative in science. Past observations *must* be repeated, not only to avoid making future mistakes, but also because that's the only way to establish new scientific facts with any confidence.[50]

In the case of psi phenomena, two types of repeatability are of interest. The first are the voluminous reports of human experience, and the second are the results of experiments testing those reports. The former provides reason to believe that something interesting is going on; the latter provides confidence about how to interpret those observations. In the next few chapters, we'll see how science has put psi to the test, and we'll discover whether these extraordinary experiences are sufficiently repeatable to reach scientifically valid conclusions.

CHAPTER 5
PUTTiNG PSi TO THE TEST

"I can't believe that!" said Alice.
"Can't you?" the Queen said in a pitying tone. "Try again: draw a long breath, and shut your eyes."
Alice laughed. "There's no use trying," she said: "one can't believe impossible things."
"I daresay you haven't had much practice," said the Queen. "When I was your age, I always did it for half-an-hour a day. Why, sometimes I've believed as many as six impossible things before breakfast."

–Lewis Carroll, *Through the Looking-Glass*

Unlike the Queen's practice before breakfast, the "impossible" facts of psi research don't require believing in anything other than measurements collected in the laboratory. These experiments come in two basic flavors: those designed to test whether information can be perceived without the use of the ordinary senses, and those that monitor the effects of mental influence at a distance. The former seems to involve information "flowing in" to the mind from the environment and, depending on how it manifests, is labeled clairvoyance, telepathy, precognition, and extrasensory perception (ESP). The latter seems to involve influence (or more likely, information) "flowing out" from the mind to the environment and is variously called mind-matter interaction, telekinesis, and psychokinesis (PK).

To test claims of perceptual psi or ESP, we isolate a test subject from a "target" object, which is hidden or placed at a distance. The target might be a photograph, a small object, or another person, and the distance can be spatial or temporal (or both). We then see if the person can successfully describe that target. To test claims of active psi or PK, we isolate the test subject from a target, which might be an inanimate object like a lightweight object, a random system like a bouncing die or radioactive decay, or a living object like a cell culture, human physiology, or human behavior. Then we see if the person can mentally influence that target to behave in a way that it wouldn't act under conditions in which no "influence" were applied. That's essentially all there is to it. Hundreds of clever variations and thousands of experiments have been conducted over the last century based on these two designs. And dozens of pitfalls have been discovered along the way that can result in false positives.

Several compilations of the "best evidence" for psi have been published since the turn of the twenty-first century.[1] They describe individual experiments that produced strong evidence for psi under unusually rigorous conditions or that introduced methods that have become enduring paradigms in psi research. Let's inspect a few examples.

DISTANT MENTAL INFLUENCE

A classic experiment in telepathy was reported in 1923 by Dr. H. I. F. W. Brugmans and his colleagues in the Department of Psychology at the University of Gröningen, The Netherlands.[2] In this experiment, a 23-year-old physics student named Van Dam was investigated for his claimed telepathic abilities. He was placed inside a curtained booth, blindfolded, and asked to place his arm under the curtain to select one square on a 6 x 8 checkerboard placed on a table next to the curtain. The target square Van Dam was attempting to select was determined randomly by the experimenter on each trial.

An assistant experimenter, called the agent, knew the target square and tried to mentally influence Van Dam's arm movements to guide him to select the correct target square. In some trials the agent was located in the same room as Van Dam; in others the agent peered through a soundproof window from the floor above Van Dam. This study was also one of the first to employ the use of a physiological measurement—galvanic skin response—to see if Van Dam's skin resistance would vary according to his selection of correct vs. incorrect targets. The results of the experiment were extremely significant, with 60 successes out of 187 trials rather than the 4 expected by chance. That's associated with odds against chance of 121 trillion to 1. There were no differences in this high performance when the agent was in the room with Van Dam or in the room above. The physiological measures provided suggestive evidence that Van Dam's skin conductance differed when he guessed correctly versus incorrectly.

A reanalysis of this study in 1978 explored in great detail a number of criticisms that had arisen over the years. It found that potential flaws, such as biases in the random target sequences and possible sensory cues, could not plausibly explain away the extremely significant results.[3] This study remains important today not only because it reported strong results under well-controlled conditions but because the use of galvanic skin response measurements spawned a continuing interest in physiologically based methods of detecting unconscious psi.

TELEPATHY

A second classic experiment that has withstood the test of time is the ESP card test, as popularized by J. B. Rhine's Parapsychology Laboratory at Duke University. This test involved cards imprinted with one of five symbols: circle, square, wavy lines, star or cross. A *deck* of ESP cards consisted of five repetitions of each symbol, for a total of 25 cards. These decks are sometimes referred to as *Zener cards* after psychologist Karl Zener, who developed the idea for such a deck. In a typical experimental *run*, the deck was thoroughly shuffled and then one person would select each card in turn and try to mentally send the symbol on that card to a distant person. This technique made it possible to collect hundreds of trials quickly, in a wide variety of environments, and under controlled conditions. Analysis of the results was straightforward, and some of the experimental results conducted under high security conditions provided extremely strong evidence for psi.

Some seem to believe that Rhine's results with ESP cards were all eventually found to be due to faulty methods, fraud, or chance. That is not true. Virtually every criticism proposed to

explain away Rhine's results, from flawed ESP cards to use of inappropriate statistics, has been discussed at great length in the relevant literature. Some have proposed that the practice of selective reporting, or publishing successful studies and filing away the unsuccessful ones, can explain the overall results of these tests. But analyses show that of the 188 experiments described in Rhine's 1940 book, *Extrasensory Perception after Sixty Years,* the combined results are so far from chance that it would take 428,000 unreported studies averaging a chance effect to eliminate the results of the known 188 experiments.[4] Given that it took 60 years to produce those 188 experiments, or about 3 studies per year, at that rate the missing studies would have taken 137,000 years to produce. I suppose it's conceivable that Cro-Magnon man was busily conducting failed ESP experiments throughout the Paleolithic Era and they failed to report any of them because writing hadn't been invented yet. But it's a stretch.

As for concerns about the quality of these experiments, philosopher Fiona Steinkamp has analyzed the Rhine-era ESP card tests in detail.[5] She found that as controls improved against such potential problems as sensory cues, recording errors, and investigator fraud the results did decline slightly, but even the most highly controlled studies had odds against chance of 375 trillion to 1 (the right most bar in Figure 5–1).

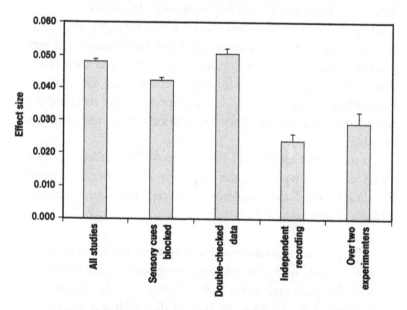

Figure 5–1. Results of ESP card tests with increasing controls. Chance expectation on this graph is an effect size of zero. "All studies" refers to 2.7 million trials recorded in 45 studies, each using the same 5-card ESP test design. The error bar at the top of each column is one standard error in length. The "true" effect, taking into account measurement errors, is at each column's height plus or minus the length of its error bar.[6]

One of the most frequently cited individual experiments from the Rhine era is the Pearce-Pratt distance telepathy test. This was conducted from August 1933 through March 1934, and consisted of 74 runs of 25-card ESP decks. The person acting as the telepathic receiver was Hubert E. Pearce Jr., a student in the Divinity School at Duke University. He had introduced himself to Rhine one day, stating that he believed he had inherited his mother's clairvoyant abilities. Rhine and his colleague Gaither Pratt had already informally conducted 700 runs of 25 cards with Pearce under a variety of conditions, and in those tests he had achieved an overall hit rate of 32% where chance expectation was 20%. This was a wildly significant outcome. As a result, they decided to conduct a tightly controlled experiment by placing Pratt and Pearce in two separate buildings.

The experiment was conducted not as a telepathy test but as a clairvoyant test. This meant that Pratt randomized and manipulated the ESP "target" cards, but he handled them face-down during each trial so he didn't know what the symbols were. Rhine also participated in this experiment by checking the recorded results and by observing portions of the test as it was running.

At an agreed upon time, Pearce visited Pratt in his research room on what was then the top floor of the Social Science Building on the Duke University campus. Here's an excerpt of Rhine's description of the test:[7]

> The two men synchronized their watches and set an exact time for starting the test, allowing enough time for [Pearce] to cross the quadrangle to the Duke Library where he occupied a cubicle in the stacks at the back of the building. From his window [Pratt] could see [Pearce] enter the Library.
>
> [Pratt] then selected a pack of ESP cards from several packs always available in the room. He gave this pack of cards a number of dovetail shuffles and a final cut, keeping them face-down throughout. He then placed the pack on the

right-hand side of the table at which he was sitting. In the center of the table was a closed book on which it had been agreed with [Pearce] that the card for each trial would be placed. At the minute set for starting the test, [Pratt] lifted the top card from the inverted deck, placed it face-down on the book, and allowed it to remain there for approximately a full minute. At the beginning of the next minute this card was picked up with the left hand and laid, still face-down, on the left-hand side of the table, while with the right hand [Pratt] picked up the next card and put it on the book . . . . This [was repeated until the pack of 25 cards was finished].

In his cubicle in the Library, [Pearce] attempted to identify the target cards, minute by minute, and recorded his responses in pencil. . . . . [Pearce] made a duplicate of his call record, signed one copy, and sealed it in an envelope for [Rhine]. Over in his room [Pratt] recorded the card order for the two decks used in the test as soon as the second run was finished. This record, too, was in duplicate, one copy of which was signed and sealed in an envelope for [Rhine]. The two sealed records were delivered personally to [Rhine], most of the time before [Pratt] and [Pearce] compared their records and scored the number of successes.

On the few occasions when [Pratt] and [Pearce] met and compared their unsealed duplicates before both of them had delivered their sealed records to [Rhine], the data could not have been changed without collusion, as [Pratt] kept the results from the unsealed records and any discrepancy between them and [Rhine]'s results would have been noticed.

In 74 planned runs, a total of 1,850 individual trials, Pearce obtained 558 hits, some 188 hits above chance expectation. This is associated with odds against chance of 10^{27} to 1. That means a 10 with 27 zeros after it, or odds of a billion billion billion to 1. After the study ended, other researchers independently examined the raw data sheets to double-check the hit rates (they matched), to see whether the sequence of trials were adequately random (they were), and to see if the results tended to cluster in bursts of hits (they didn't), and a variety of other ideas. In short, the overall picture indicates that Pratt was able to do what he said he could do: describe cards at a distance. Rhine's comment about this experiment was circumspect: "The series contributed all that an experiment can do toward establishing the ESP hypothesis. The rest is a question of receptivity on the part of the professional group."

This study involved a single talented participant, so the outcome does not generalize to the population at large. Pearce's outstanding performance meant that in the typical run of 25 cards, rather than obtaining about 5 hits on average, he obtained 7.5 hits. That doesn't sound very impressive, but because Pearce sustained that performance over 1,850 trials we know to very high levels of certainty that those 2.5 extra hits per run were not due to chance.

A few months after the conclusion of this experiment, Pearce lost the high scoring ability he had displayed for the prior two years.[8] In fact, many of the high-scoring individuals in ESP card tests eventually lost their abilities, some after a few thousand runs and some after tens of thousands. The most obvious reason for such declines is that these tests are exciting, fun, and motivating for about 10 minutes. Then it's like an anesthetic wearing off before surgery is finished. The test becomes more and more painfully boring until eventually you'd rather poke a stick in your eye than continue to guess cards. This increasingly antimotivational factor eventually extinguishes the very skill of interest.

Still, the results of this test demand an explanation, and we know that chance is ruled out. So what remains? When all else fails, the explanation of last resort is always fraud. Psychologist Mark Hansel, in his 1964 book *ESP: A Scientific Evaluation,* proposed that Pratt did not stay in the library as planned. Instead, said Hansel, Pearce secretly made his way back to the building where Pratt was located. Then he went into a room across the hall from where Pratt was located (Hansel didn't know that particular room was being used by students at the time of the experiment), and then Pearce supposedly stood on a chair near a door and looked down through a transom in the door into Pratt's office, where he watched him record the sequence of cards. To bolster his explanatory scenario, Hansel included a diagram of the rooms as he remembered them during his visit to the Duke Laboratory in 1960. His diagram's caption included the words, "not to scale," because when Hansel went to the Duke architect's office he was unable to obtain the floor plans for that building for the period in the 1930s when the tests were conducted. If Hansel *had* obtained the actual floor plan, he would have seen that his "peeking hypothesis" was impossible.[9] Still, this experiment, like any single experiment with exceptionally good results, will never be able to convince those who are stubbornly incredulous.

CLAIRVOYANCE

In a picture-drawing psi experiment, one person selects or imagines an object and sketches it, and then concentrates on "sending" that image to a distant partner. The partner attempts to reproduce the same object or sketch, and then the two pictures are judged for similarities.

Some of the earliest picture drawing tests would not satisfy modern standards of experimental controls because the target images weren't always selected completely at random. We now know that people are quite poor at randomly selecting things,

and thus one person's "random" sketch might closely correspond to a partner's "random" drawing because of shared memories and experiences. Say a couple decided to conduct such a test after spending the day at the ocean. Water themes would likely be bubbling in the back of their minds, so an image of say, a seagull, would be far more likely to spontaneously arise in both of their minds than say, a cactus, and this shared memory could easily mimic a case of telepathy.

With that caveat in mind, some of the early picture-drawing experiments were still quite impressive. In addition, researchers at the time were aware that shared biases were a possible loophole, so they experimented with different pairs of people and different methods of selecting targets to see if such factors made a difference in the results.

UPTON SINCLAIR

An example of a particularly successful series of picture-drawing experiments is reported in the book *Mental Radio,* published in 1930 by the American social activist Upton Sinclair. Sinclair first rose to fame because of his novel, *The Jungle* (1906), which described the horrendous working conditions and lack of sanitation in the Chicago meatpacking houses. That book led to the U.S. government's Pure Food and Drugs Act and the Meat Inspection Act (both in 1906). Many years later, Sinclair's novel *Dragon's Teeth* (1942), about the rise of the German Nazi movement, won him the Pulitzer Prize.

Sinclair wrote *Mental Radio* in collaboration with his wife, Mary Craig Sinclair. "Craig," as she was known, was the principal talented subject of the book, and it was her repeatedly demonstrated skills that finally convinced the skeptical Sinclair, and many others, that telepathy exists. In these experiments, first Sinclair sketched a small object, then Craig, located at a distance, would try to mentally perceive the sketch and reproduce it. Sometimes a family friend would make the sketch, and some of the tests were conducted with Craig many miles away from the person drawing the picture. *Mental Radio* reproduces dozens of examples of their tests, showing striking similarities far beyond what one would expect by chance.

Craig finally convinced Sinclair that he should put his convictions on record in the form of a book, even though—ironically—telepathy was far more controversial than Sinclair's promotion of social justice. As he put it, "There isn't a thing in the world that leads me to this act except the conviction which has been forced upon me that telepathy is real, and that loyalty to the nature of the universe makes it necessary for me to say so."[10] Sinclair's friends later wrote review articles with barbed titles like "Sinclair Goes Spooky." But he stuck with it, explaining that "It is foolish to be convinced without evidence, but it is equally foolish to refuse to be convinced by real evidence."[11]

Albert Einstein was one of Sinclair's many prominent friends. Einstein was skeptical about telepathy, but he trusted Sinclair's integrity and he was willing to consider his data carefully. After reading the book, Einstein agreed to provide a preface for the German translation of *Mental Radio*.[12] Einstein wrote the following:

> I have read the book of Upton Sinclair with great interest and am convinced that the same deserves the most earnest consideration, not only of the laity, but also of the psychologists by profession. The results of the telepathic experiments carefully and plainly set forth in this book stand surely far beyond those which a nature investigator holds to be thinkable. On the other hand, it is out of the question in the case of so conscientious an observer and writer as Upton Sinclair that he is carrying on a conscious deception of the reading world; his good faith and dependability are not to be doubted. So if somehow the facts here set forth rest not upon telepathy, but upon some unconscious hypnotic influence from person to

person, this also would be of high psychological interest. In no case should the psychologically interested circles pass over this book heedlessly. [signed May 23, 1930]

Some of Sinclair's friends begged him not to publish the book for fear of ruining his reputation. One friend said that the results could not possibly be true, for real telepathy would require him to abandon "the fundamental notions on which his whole life has been based."[13] Fear is a common response when one's basic beliefs are challenged.

RENÉ WARCOLLIER

A second example of picture-drawing experiments is described in the book *Mind to Mind,* published in 1948, by French researcher René Warcollier. Much of this book's contents were presented by Warcollier as an invited lecture to the Sorbonne, one of the most prestigious universities in Europe.[14] Warcollier's earlier book, *La Télépathie,* published in 1921, was a sensation in France, and was reprinted in English with additional material in 1938 under the title *Experimental Telepathy.*

Warcollier was already convinced that telepathy existed through the work of Rhine and others, so his books primarily explored how it worked. Most of his experiments and analyses focused on ways that the original target images were distorted or otherwise misperceived by the recipients. He noted that images were not transmitted like photographs but were "scrambled, broken up into component elements which are often transmuted into a new pattern."[15]

What Warcollier demonstrated is compatible with what modern cognitive neuroscience has learned about how visual images are constructed by the brain. It implies that telepathic perceptions bubble up into awareness from the unconscious and are probably processed in the brain in the same way that we generate images in dreams. And thus telepathic "images" are far less certain than sensory-driven images and subject to distortion.

WHATLEY CARINGTON

A third picture-drawing experiment was reported in 1941 by Cambridge University psychologist Whatley Carington, who referred to these studies as experiments in "paranormal cognition." Carington pioneered the use of random selection of drawings and cross-matching statistical analyses. He also employed a third-party investigator to ensure that the data were properly handled and recorded, and to provide protection against fraud. Carington was motivated to conduct his experiment because he was persuaded that previously reported picture-drawing experiments provided sound evidence for psi. As he wryly put it,

> [Those] studies convinced me that, despite the machinations of the malevolent hoodoo which apparently dominates the subject, the case for supposing that significant and genuine positive results had been obtained in the past from experiments of this kind was very strong.[16]

Carington set out four principal goals: (1) The design and conditions had to be specified clearly in advance, (2) the scoring must be unbiased, (3) the results had to be statistically significant, and (4) the experiment must be repeatable. Each of his resulting picture-drawing experiments lasted ten nights, one drawing per night. On each of those nights Carington or his wife would make a drawing. The topic of the drawing was determined by opening a book of mathematical tables at random, selecting the last digits of the first three or four items in that book, and then opening to that page of *Webster's* dictionary and finding the first drawable word on or after that page. Carington then posted a drawing of that word on a wall in his home study at 7 p.m., and left it there until the next morning. He describes this room in detail and the precautions he took to prevent fraud via peeking through windows or by someone entering the room.

The participants in his experiments attempted to perceive the drawings wherever they happened to be and at any convenient time while the drawing was posted. There were a total of 5 experiments, each with 10 drawings, for a total of 50 targets. About 250 people took part in the study as percipients; together they produced 2,200 drawings. Fewer than a dozen people took part in more than one experimental series.

An independent judge who was unaware of the target drawings evaluated the participants' sketches. By cross matching the original sketches against the participants' drawings, he found 1,209 drawings to be similar to the targets. From this blind matching method, Carington calculated how many "hits" on the targets might be expected by chance, and then compared this to the number of hits actually obtained. From this exercise, he concluded the following:

> It is found that the excess is such as would be equaled or surpassed only about once in some thirty thousand such investigations if chance alone were responsible. In other words, percipients' drawings resemble the originals (considered as a group) at which they are aiming more closely than they resemble originals at which they were not aiming to an extent which cannot plausibly be attributed to chance. [17]

Carington then added something that modern psi researchers would agree with:

> It seems to me that what I have found . . . is eminently compatible with both sides of common experience—with the knowledge that on the whole people very seldom show signs of paranormal cognition, and with the knowledge that none

the less they occasionally do. Finally, the fact that . . . the ability concerned is pretty widely distributed; or at least not concentrated to any startling degree among a very few specially gifted persons, suggests that it is likely to prove an attribute common to all humanity, with nothing alarmingly magical about it; so that perhaps the adjective "paranormal" is something of a misnomer after all.[18]

REMOTE VIEWING

Remote viewing was a term coined by physicists Harold Puthoff and Russell Targ at SRI International in the early 1970s. It refers to a task of clairvoyance in which one person (the "agent") travels to a randomly selected distant location while the remote viewer, secured in a laboratory, describes where the agent went.[19] An article about these experiments, published in *Nature* by Puthoff and Targ, attracted criticism (of course), but detailed examinations of the critiques found them unable to explain away the reported results.

Targ, physicist Edwin May, and many others have since replicated remote viewing under rigorously controlled conditions many times.[20] Princeton University's Engineering Anomalies Research (PEAR) Laboratory generated one of the single largest remote viewing experiments.[21] In a 2003 report, former Princeton University Dean of Engineering Robert Jahn and psychologist Brenda Dunne summarized 25 years of remote viewing (they call it remote perception) research.[22] They conducted 653 formal trials from 1976 to 1999, involving 72 participants. Most of those trials were conducted precognitively, meaning the future target was randomly selected *after* the percipient had recorded his or her impressions.

The PEAR Lab developed increasingly refined analytical methods over the years, significantly expanding the simple methods used half a century earlier by Carington. Their goal was to develop quantitative ways to measure the similarity between the remote viewers' impressions and the agent's experiences. Their overall assessment of the matches in the 653 trials provided strong evidence that the results were definitely not due to chance (odds against chance of 33 million to 1). As Jahn and Dunne wrote, "The overall results of these analyses leave little doubt, by any criterion, that the [precognitive remote perception data] contain considerably more information about the designated targets than can be attributed to chance guessing." Further analyses indicated, as also noted earlier by Puthoff and Targ and many previous researchers, that the remote perception outcomes appeared to be independent of both distance and time.

"Yet, like so much of the research in consciousness," Jahn and Dunne then noted, "related anomalies, replication, enhancement, and interpretation of these results proved elusive. As the program advanced and the analytical techniques became more sophisticated, the empirical results became weaker."[23] This doesn't mean that the remote perception results declined as the *controls* were tightened, because the experimental design was rigorously controlled throughout. Instead, as the analytical methods became increasingly focused on extracting the "signal" from the "noise," the signal began to disappear. Jahn and Dunne speculated that this might mean that the signal might actually *require* some noise. This is analogous to a paradoxical physical phenomenon known as stochastic resonance, in which an increase in noise enhances the detection of weak signals. For example, patients with reduced sensitivity in their feet often find it difficult to maintain their balance while standing or walking. Given this, one might think that standing on mechanically vibrating insoles would further decrease foot sensitivity, but in fact the opposite occurs—sensitivity and balance improve.[24] The additional noise from the vibration boosts weak sensations from the feet and makes them easier to sense. Similar stochastic resonance phenomena have been found in the sensory systems of many living systems.

*   *   *

As we've seen, some individual experiments and results reported by individual laboratories have produced strikingly successful results. But however good those experiments may be, they tend not to be convincing to other scientists. There's always the suspicion that the investigators may have made some sort of mistake, or worse, perpetrated fraud. This is why science values independent replication. It's unlikely that different investigators will repeatedly make the same mistake, or secretly conspire to fool everyone. This provides the motivation for examining analyses of *collections* of experiments, known as meta-analysis.

CHAPTER 6
CONSCiOUS PSi

"So I wasn't dreaming, after all," she said to herself, "unless–unless
we're all part of the same dream. Only I do hope it's *my* dream, and
not the Red King's! I don't like belonging to another person's dream,"
she went on in a rather complaining tone: "I've a great mind to go
and wake him, and see what happens!"

–Lewis Carroll, *Through the Looking-Glass*

For many years researchers tried to develop the Holy Grail of
psi–the easily repeatable experiment. Any high school student
should be able to conduct this overwhelmingly convincing experiment, like demonstrating the effect of gravity by dropping a
rock. Ideally, no form of personal judgment or evaluation
would be required, so anyone could immediately see the results
as self-evident. That quest has, so far, eluded our best efforts.
This frustration had led some to believe that such an experiment is impossible, and hence that claims for psi lie outside the
bounds of science.

But the key word is *easily* repeatable. As we'll see, psi experiments are indeed repeatable. They're just not trivially easy to
repeat. For that matter, hardly anything involving skilled
human performance is absolutely predictable, except perhaps
stubbornness in the face of evidence one doesn't wish to see.

Why are spectator sports are so wildly popular? Precisely because of the uncertainty in how each player will perform. Given the known variations in highly skilled performance, you can imagine the range of variations observed in *average* players. The same is true for psi performance. Most psi experiments have been conducted with average people claiming no special skills, and not surprisingly, the results have shown large variations in performance.

To explore this important question of repeatability in psi experiments, over the next few chapters we'll review more than a thousand experiments. Taken together, we'll find that these studies provide repeatable, scientifically valid evidence for psi. This ought to immediately arouse your inner skeptic. You ought to be thinking, "Oh yeah? Show me." We'll get to that in a moment. But first a word about a dilemma that arises when evaluating this sort of evidence.

It's often said that "the devil's in the details." This means it's easy to make sweeping and dramatic claims, but when you pay close attention to the underlying details you often discover that the claims weren't as strong as you've been led to believe. To use a distressing political analogy, when President George W. Bush decided to invade Iraq in 2003, he "sold" the war on the basis of the Central Intelligence Agency's (CIA) claims about Iraq's weapons of mass destruction. As it turned out, there were no such weapons. A commission subsequently formed by President Bush to review the CIA's mistake concluded that the claims weren't slightly wrong, they were "dead wrong."[1] This highlights one of the more dramatic consequences of overlooking details.

So the devil is indeed in the details, but unfortunately those very same details often involve technical jargon and concepts that can be devilishly tricky to understand without specialized training. This is especially true when it comes to assessing claims about controversial experiments. So my dilemma is to demonstrate that the evidence we'll be discussing is not "claims of mass seduction," but at the same time not get so bogged down in minutiae that this book becomes the world's most efficient cure for insomnia. My solution in wielding this double-edged sword is to place most of the technical bits and journal citations in the endnotes. If you want to keep it high and dry, don't go there. If you're passionate about nitty-gritty details, now you know where to find them.[2]

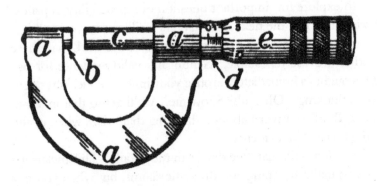

MEASUREMENT ERROR

All measurements involve some error. This is as true in physics as it is in psychology. Measurements involving human behavior are especially uncertain because we aren't rocks. We are exquisitely sensitive and dynamically reactive to the environment; our behavior is modulated by how we feel physically and emotionally, and we adjust our behavior depending on what we think others want to see. Laboratory methods impose further artificial constraints so that only certain limited forms of behavior can be measured reliably. This means that psi might *appear* to be weak and erratic, but that's at least partially because the tools being used to study it are inappropriate. Like trying to catch a one-inch fish in a two-inch net, most of the time your quarry is going to escape no matter how meticulously you wield the trap.

Nevertheless, if we're going to be scientific about this we're obliged by the rules of evidence to conduct controlled laboratory experiments, and thus we're forced to deal with problems like measurement noise and a pale reflection of how psi appears spontaneously in real life. What we gain in return is high confidence that the effects we observe aren't due to a litany of common explanations, like coincidence or reactions to ordinary sensory information.

As soon as we accept that experimental results aren't going to be perfect, we have to deal with questions like, "How good does performance have to be," to reject explanations like coincidence. And "How many times do we need to repeat an experiment," to make a convincing case that something interesting is going on. The usual answer to these questions is the skeptic's mantra: "Exceptional claims require exceptional evidence."[3] But how do we know when evidence is sufficiently exceptional? In principle, as more and more evidence accumulates, eventually the data ought to exceed some threshold of persuasion that would overcome any degree of skepticism. Unfortunately, in the real world this strategy doesn't work.

In the 1980s, I worked on a top secret psi research program for the U.S. government (now declassified). At the first research briefing I attended, I was shown examples of high-quality remote viewing obtained under exceptionally well-controlled circumstances. I asked in amazement, "Why is psi still considered controversial by the scientific mainstream? Why not just conduct an experiment of 20 or 30 trials with this type of remote viewing skill? That ought to convince anyone that psi is real." The answer, explained to me patiently by physicist Ed May, was simple. He said, "You're making the 'rational man' mistake." He meant that we usually assume that science is a rational process, but it's not. When we're presented with evidence that counters our prior beliefs, instead of the new evidence swaying us *toward* a new or revised belief, it tends to reaffirm our *prior* beliefs. Well, I thought, that's completely ridiculous. It's got to be a mistake. Unfortunately, after witnessing precisely these reactions to the data for twenty years, I have reluctantly concluded that the "rational man" hypothesis is indeed false.[4]

The technical term for one form of this irrational phenomenon is the "confirmation bias." This psychological quirk causes evidence supporting your beliefs to be perceived as plausible, and evidence challenging your beliefs to be perceived as implausible.[5] Studies in social psychology have repeatedly demonstrated that journal reviewers invariably judge articles being submitted for publication according to their prior beliefs. Those who agree with a hypothesis tend to judge a paper reporting positive results as an excellent piece of work, and those who disagree judge the very same paper as a flawed failure. The former referees recommend publication and the latter don't. The final decision to publish is left up to the editor, so if the editor doesn't happen to agree with the paper's hypothesis then there's a good chance it won't appear in the journal. And then the evidence doesn't exist as far as the rest of the scientific community is concerned. In science, this tends to create a genteel "good old boys" club of acceptable ideas, while unacceptable ideas are consigned to the biker's bar lounge on the wrong side of the tracks. Fortunately, most scientists also tend to have high curiosity, so the club's rules can change with sufficient persistence (and after the retirement of some of the older good old boys).

META-ANALYSIS

The confirmation bias can be overcome in three ways: A practical application can be developed, a testable theoretical explanation can be confirmed through new experiments, or consensus opinion can be swayed by authority. None of these are likely to occur without first being able to demonstrate that the effects are *independently repeatable* under laboratory conditions. How do we demonstrate repeatability? We analyze the outcomes of *collections* of previous experiments, or said another way, we analyze previous analyses. This is called *meta-analysis*.[6]

Meta-analysis has become an essential tool in the so-called soft sciences, including ecology, psychology, sociology, and medicine. Thousands of meta-analyses have been published, there are scientific journals devoted to meta-analyses, and it has become the basis for the new emphasis on evidence-based medicine. Because of the increasing importance of meta-analysis, the methods used to combine experimental results have steadily improved over the years.

Three questions are often asked about meta-analysis: First, how can the act of combining different experiments tell us whether an effect is repeatable? Second, what if some of the experiments you're examining are well designed, but others aren't? And third, what if the studies you've found are just the successful ones, and you're ignoring the failed experiments? These questions are known as the *apples-and-oranges problem*, the *quality problem*, and the *file-drawer problem*.

The apples-and oranges-problem asks if it's valid to combine studies using different investigators, designs, and subjects. The answer is yes if we wish to learn what's common in apples and oranges, namely, something about the nature of *fruit*. When a series of psi experiments are combined, apples and oranges are represented by unavoidable differences among the studies, but the effect in common—the psi effect—remains the same. Of course, if you're only interested in red fruit, or if you're specifically uninterested in cherries, then you have to be clear about this in your selection of fruits to combine. In the meta-analyses discussed here, I'm interested in a very broad, proof-oriented assessment of the evidence, so I've included as many relevant studies as possible.

The next question asks if it's valid to combine experiments of varying quality, referring to how well an experiment is designed and executed. A sloppily conducted experiment shouldn't carry as much evidential weight as a carefully conducted one. In general, if increases in study quality are associated with smaller effects, that signals a potential problem. Many ways of assessing experimental quality have been developed.[7]

The third question addresses the fact that investigators tend to publish studies with positive outcomes, but they don't publish studies with negative outcomes. This is known as the selective reporting or *file-drawer problem,* so called because of supposed stacks of unsuccessful studies languishing in researchers' backroom file drawers. If there are lots of unpublished, negative studies, this missing data can nullify the published evidence. Methods for estimating the size and the effects of the file-drawer problem have become increasingly sophisticated in recent years.

Finally, a word about the type of meta-analyses discussed here. There are two major types: proof-oriented and process-oriented. The goal of the first type is to see *whether psi exists.* The goal of the second type is to find out *how it works.*[8] We'll focus on meta-analyses of the first kind because there's no sense in worrying about how something works before you're reasonably sure that it exists. With that as a brief introduction to our approach, let's see what happens when we apply meta-analysis to the question of psi in dreams.

PSI IN DREAMS

Mrs. Anne Ring sent me the following letter recounting her psi experience in a dream:

Many years ago I had a very strange dream concerning my father. I dreamt that he was decorating the house (the way we do in England—or used to—with paper chains, holly, etc.). Except the decorations he was using were not the type used for Christmas. Suddenly he sat down on a chair and collapsed and he died. I woke up crying so loudly that it woke up my husband. I looked at the clock and it was exactly 2 a.m. California time. I told my husband the dream and he just said, "Well it's nothing, you are always having strange dreams, go back to sleep." But the dream had disturbed me and it took a long while for me to get back to sleep.

The following morning was Thanksgiving Day and as I was preparing the meal the telephone rang and it was my brother calling from London to say my father had died. It was a terrible shock because I had seen him in the May of that year and he was in robust health (in fact, he had not ever been ill or in hospital in his life). I asked my brother when it had happened and he replied that our stepmother had just called him and told it had happened at 10 a.m. London time. The exact moment that I had the dream (2 a.m. California time). By the way, he was putting up decorations because it was his wedding anniversary to my stepmother and they were going to have a party that night.[9]

How shall we interpret this experience? Is it a poignant coincidence or is it a case of genuine clairvoyance? This was the one and only time Mrs. Ring ever had a dream like this, and it contained details and timings that matched real-world events. I've been told similar experiences by professors at major universities, by program directors at the NSF, and by generals in the Army. These are not naïve people prone to fantasy. They appreciate the difference between meaningless coincidence and genuinely exceptional events.

One possible explanation about such stories is that given the billions of nightly dreams around the world, surely some of them will occasionally come true by chance. We'd hear about those dreams, and then we'd imagine that psi in dreams must be quite common. And indeed, cross-cultural surveys do show that about half of all spontaneous psi experiences occur in dreams, and many of them involve accidents or the death of a distant family member.[10] Because of the frequency of these reports, researchers became interested in seeing whether similar psi experiences could be evoked in controlled laboratory settings where the most obvious explanation—coincidence—could be strictly evaluated.

One of the first such tests occurred in 1960, when psychiatrist Montague Ullman tested medium Eileen Garrett. Garrett was president of the New York City–based Parapsychology Foundation and a philanthropist who helped fund many scientific investigations of psi experiences. With Garrett's support, Montague Ullman, psychologist Karlis Osis, and engineer Douglas Dean established a sleep laboratory at the Parapsychology Foundation in New York. On June 6, 1960, Dr. Osis selected a pool of three pictures from *Life* magazine, sealed them in envelopes, and gave them to Garrett's secretary to take to her home some miles away. She was to wait for a phone call from the foundation indicating that Garrett had fallen asleep in the lab, and then to randomly shuffle the envelopes, select one of the envelopes, and telepathically communicate the image to her.

Dean and Ullman stayed up all night to monitor Garrett, but they were disappointed because they saw no signs that she had gone into the rapid eye movement (REM) state indicative of dreaming. As a result, they didn't phone the secretary to start sending. But Garrett did have a dream that night about horses furiously running uphill. She said later that the dream reminded her of a chariot race from the movie *Ben-Hur,* which she had seen two weeks before.[11] Ullman later learned that one of the target pictures was a color photo from *Life* magazine of the chariot race from *Ben-Hur*. This was unexpected but intriguing, and so Ullman moved the program to his sleep laboratory at Maimonides Medical Center, in Brooklyn, New

York. After running a series of pilot tests, he started conducting formal trials.

From 1966 through 1973, Ullman, psychologist Stanley Krippner, and many coworkers ran a total of 379 dream psi sessions. In most of these sessions, a volunteer receiver (say Jill) spent the night in the Maimonides dream lab. Jill met and talked with an experimenter (Jack), who acted as the sender. Jill also met the other investigators taking part in the testing session that night.

When Jill was ready to go to sleep, she was ushered into a soundproof and electromagnetically shielded room. Such chambers are commonly used in rigorously controlled psi experiments to ensure that the participants aren't responding to any ordinary signals. Once in the chamber, an experimenter applied electrodes to Jill's head to monitor her brainwaves and eye movements. From that point on she had no further contact with Jack or anyone else until the session was completed. In a room next to the experimental chamber, a technician monitored Jill's brainwaves and eye movements throughout the night. When the technician observed that rapid eye movement (REM) began, and thus dreaming had begun, Jack was notified.

In some of the Maimonides studies, Jack was located in a room about 32 feet from Jill. In other studies, Jack and Jill were 98 feet, 14 miles, and in one case, 45 miles apart. Before Jack left for his remote location, an assistant gave him a sealed envelope containing a picture that had been randomly selected from a pool of possible pictures, usually a pool of 8 or 12 pictures. When Jack arrived at the remote site, he opened the envelope. During the experiment, only Jack knew which picture had been selected. To ensure that no one else could accidentally figure out the target during an experimental session, the only form of communication allowed between Jack and the experimenters was a buzzer tone or series of planned telephone rings. Each time Jack received one of those signals, he tried to mentally influence Jill's dream based on images from his picture.

When Jill stopped dreaming, another signal was sent to Jack to tell him to stop sending. Then a lab technician woke up Jill and asked her to describe the dream that she just had. After audio taping the dream, she was allowed to go back to sleep. After each successive dream over the course of the night, she was reawakened and the reporting process repeated. This happened three to six times over the course of a typical night's sleep. In the morning, Jill was roused again and asked for her overall impressions of the picture that Jack was trying to send. Her responses were recorded and transcribed for later analysis.

To evaluate Jill's impressions in each dream session, one or more independent judges later examined the transcript of her dreams and compared it to the full pool of pictures, one of which was the image that Jack was trying to send to Jill. The judges, who didn't know the real picture (called the target), were asked to provide a *ranking* of how well each picture matched Jill's dreams. So the picture with the highest correspondence to the transcript would be ranked one, and the picture with the least correspondence would be ranked, say, eight, assuming the pool had eight pictures. If the judge ranked the target picture in the top half of the pool, from rank one through four, this was considered a hit, otherwise it was a miss. If dream psi was really just due to coincidences, then over many repeated sessions the hit rate in this experiment would be like a coin flip: it would fluctuate around the chance-expected rate of 50%.

Advice to self-testers: The participants in these studies knew what to expect in advance and were willing to give up a good night's sleep for the sake of science. If you think it might be fun to stare at a sibling, parent, or spouse all night and repeatedly wake him or her up to see if your thoughts are showing up in his or her dreams, it's probably a good idea to get their permission first. Fortunately, there's an easier way to conduct dream psi experiments that doesn't require having to wake anyone up.

The new method was designed because it took the Maimonides lab seven years to complete 379 dream psi sessions.

That averages to a single data point per week. To speed up the process, a new generation of investigators took advantage of the fact that people dream every night and most people can learn to remember their dreams. In an at-home dream psi experiment, a computer is programmed to automatically select a picture from a random pool of pictures and to display that target image repeatedly during the night, on a computer monitor in an empty room, usually between 3 and 4 a.m. The computer is in a locked room at a distant location with no one present and shielded from view to prevent anyone from peeking through windows or otherwise figuring out what the target is.

Each participant keeps track of his or her dreams at home, then the next morning at the lab all participants view four pictures: the actual target and three decoys. They individually rank the four images according to how well each one matched their dreams, and then the ranks are combined to create a single consensus vote on the best possible match. Only the computer knows the actual target that it displayed during the previous night, and after the vote is cast the participants get to see if their choice is correct. This allows one session to be collected per night, but unlike the dream lab studies it doesn't require a special laboratory, or all-night technicians, or even separate senders or judges.[12]

META-ANALYSIS

In 2003, British psychologists Simon Sherwood and Chris Roe from University College Northampton, England, reviewed all of the dream psi studies from the original Maimonides series through the latest at-home dream experiments.[13] All of these tests shared two key factors: They all tested whether information at a distance could be perceived in dreams, and they were all conducted under controlled conditions that excluded mundane explanations like sensory cues or recording mistakes.

Sherwood and Roe found 47 experiments involving a total

of 1,270 trials. The overall hit rate was 59.1% where 50% is expected by chance (Figure 6.1). This 9.1% increase over chance may not sound like much, but it's associated with odds against chance of 22 billion to 1. That rules out coincidence as a viable explanation.

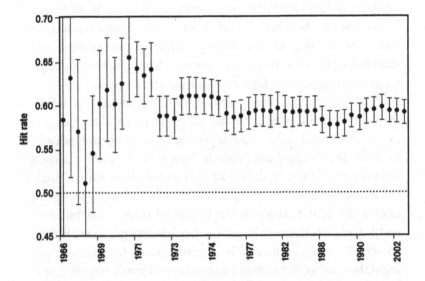

Figure 6–1. Cumulative average estimate of the hit rate in all known dream psi experiments, from 1966 to 2004, with one standard error bar. Chance expectation is at 50%. The overall estimate is associated with odds against chance of 22 billion to 1, so coincidence is not a plausible explanation for the results of experiments studying psi in dreams.

Because no measurement is ever absolutely correct, "error bars" are often used on graphs to show the range in which the true effect is believed to reside. Error bars allow us to see at a glance how close the hit rate is to chance expectation. In this case, the dream psi effect is 6.4 "standard errors" above chance. This provides 99.999999996% confidence that the results exclude chance. So we know, with high confidence, that Jill's dream content matched what Jack was mentally sending or a distant computer was displaying.

If not chance, then what else could explain these results? One possibility is that the experiments were poorly designed, and so all we're seeing are the effects of errors or flaws. But reading the actual experimental reports quickly reveals that such an explanation is implausible. The researchers who conducted these studies were well aware of the many pitfalls that can contaminate experiments, and these studies were specifically designed and executed to avoid those problems. More formal methods of evaluating the effects of varying experimental quality have been conducted, and they confirm that the dream psi studies were not due to poor designs.[14]

So maybe the strong statistical results are due to the fact that we're only considering the positive studies and ignoring the failed ones. If thousands of studies were accidentally overlooked, or failed to be reported, and all of those "missing" studies provided no evidence for psi, then our odds of 22 billion to 1 could be a vastly inflated figure. To see if this is the case, we have to answer two questions: First, assuming that we've missed some failed studies, how many would be required to nullify the results that we've observed? If that figure is small, then we'd have to conclude that the evidence for psi in dreams isn't so good after all. And second, is there a way to estimate if any studies are actually missing? These two questions about the "file drawer" of potentially missing studies will be encountered repeatedly as we review the experimental evidence, so it's worth examining in a bit more detail how they're answered.[15]

EXPLORING THE FILE DRAWER

For the dream psi experiments it turns out that an additional 700 studies, averaging an overall chance outcome, would be needed to bring the observed results down to chance.[16] Considering that about 20 different investigators have reported dream psi studies, this would mean that each of those investigators would have had to conduct, but not report, 35 failed experi-

ments for *each* experiment with a positive result that they did report. Given that the average dream experiment involved 27 sessions, these 700 supposedly missing experiments would imply that about 700 x 27 or 18,900 sessions were conducted but not reported. One dream session takes one night, so we'd have to conclude that 18,900 nights, or over 50 years' worth of data, wasn't reported. That hardly seems plausible.

A more conservative approach assumes that the missing studies don't average out exactly to chance, but rather to a slightly negative effect because of asymmetries in the technical meaning of a statistically "significant" experiment.[17] Based on this assumption, the number of studies needed to wipe out the overall results is reduced to 670. This works out to 49 years' worth of missing data.[18] We can conclude from this that the file-drawer problem is not a plausible explanation for the results of the dream psi studies.

There's another way to test whether selective reporting might be a problem. It's called the funnel plot because of the shape of the plot. The inverted funnel occurs because studies with few repeated samples produce less accurate measurements than studies with lots of repeated samples. So when we plot an experiment's sample size (the number of repeated measurements) versus a measure of its effect size (the magnitude of the overall results), we'll end up with a nice

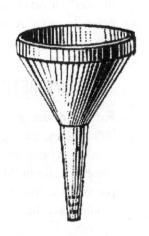

symmetric funnel shape centered around some average value—provided there's no selective reporting problem (Figure 6–2). That average value is the best estimate of the effect we're interested in.

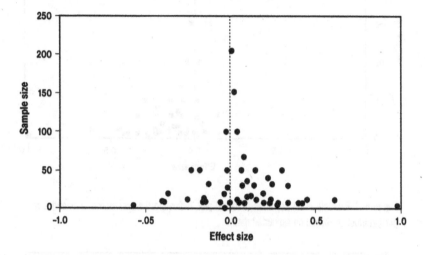

Figure 6–2. Example of a symmetric funnel plot with no selective reporting problem (based on simulated data).

When selective reporting takes place, the studies most likely *not* to be published are those with smaller sample sizes and negative outcomes. That's because smaller studies tend to be pilot tests, and negative pilot tests are easy to "forget" to publish. The consequence of this reporting bias is that the lower left side of a funnel plot will appear to have a "bite" taken out of it (Figure 6–3). There's no missing bite in the funnel plot for the dream psi studies, so there's no evidence of selective reporting (Figure 6–4).[19]

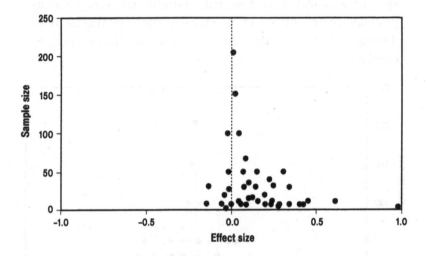

Figure 6–3. Example of an asymmetric funnel plot, indicating the presence of a selective reporting problem (based on simulated data).

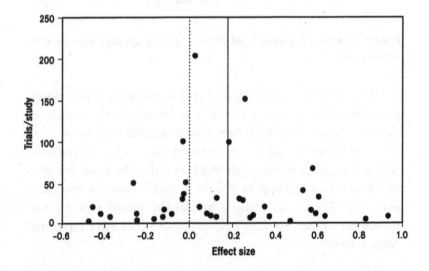

Figure 6–4. Funnel plot for the 1,270 dream psi sessions. The solid vertical line shows the overall average effect observed in these studies; the dashed vertical line shows chance expectation by comparison.[20] There is no selective reporting problem.

What we've learned so far is that the combined results of the dream psi studies aren't due to chance, the studies were carefully designed to avoid all known flaws, and the overall assessment isn't affected by selective reporting biases. Could the results be due to fraud? Participant fraud is ruled out by the experimental designs, which are controlled to prevent both accidental and intentional cues. And investigator fraud isn't viable because independent groups have successfully repeated these studies for more than three decades. Not every experiment is successful, but overall it's clear that something interesting is going on.

What remains is the impeccable logic of Sherlock Holmes, who reasoned that once all other factors are plausibly eliminated, what remains must be true. In this case, the remaining truth is that to very high levels of confidence we know that information at a distance was successfully perceived in dreams under well-controlled conditions. But let's remain cautious. If this conclusion is correct, then it ought to be possible to detect similar psi effects in states of awareness that closely resemble dreaming. Let's see if this is so.

PSI IN THE GANZFELD

Ganzfeld is a German word meaning "whole field." It's a mild form of sensory stimulation originally developed by gestalt psychologists to study the nature of visual imagery.[21] In a ganzfeld psi experiment the participant, say Jill, is asked to relax in a comfortable, reclining chair. The experimenter places halved Ping-Pong balls over her eyes and gives her headphones to wear that play pink noise, a whooshing sound like a deep throated waterfall.

Then the experimenter shines a red light on Jill's face, and she is asked to keep her eyes gently open under the Ping-Pong balls. All she'll see is a soft red glow everywhere she looks. Soon she won't be able to tell if her eyes are open or closed, and this, combined with the whooshing sound she's hearing, will eventually stimulate her brain to provide something more interesting. Many people in the ganzfeld condition describe that a pleasant, dreamy state of awareness is evoked within a few minutes.

After being allowed to relax in this dreamlike reverie for 15 minutes, Jill is asked to speak aloud (the jargon is to "mentate") anything that comes to mind over the next 30 minutes, while Jack, at a distance, tries to mentally "send" an image to her. In most ganzfeld setups, Jill's mentation is audio recorded, and in some of the newer setups Jack's target imagery (a picture or a one-minute video clip) is recorded along with Jill's mentation to allow independent judges to later examine Jill's impressions in comparison to the target image that Jack was viewing.

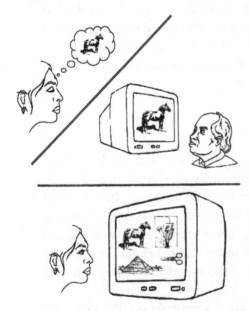

Figure 6–5. Illustration of ganzfeld telepathy experiment design. Above: Jack mentally "sends" a picture to Jill, who is at a distance and imagining what Jack is viewing. Below: After the sending period Jill attempts to match her mental impressions with one of four pictures, one of which was the real "target" image, along with three decoys.

Over the course of a 30-minute session, Jack views the same photograph or a repeatedly played video clip (Figure 6–5). This target image is randomly selected from a pool of four possible images. Each of these pools is formed so that each of the four images within a pool is as different from each other as possible. A typical pool might consist of say, a one-minute video clip of a desert scene, a second video clip of a city scene, a third involving a person eating an ice cream cone, and fourth involving a fish swimming in the ocean. A computer randomly selects one of those clips, and Jack is asked to mentally send that target imagery to Jill.

In most modern ganzfeld designs, Jack can listen to Jill over a one-way audio link as she describes her ongoing imagery. In this way Jack can use Jill's impressions to help adjust his mental sending strategy, like a kind of biofeedback. During the 30-minute sending period Jack might send the video clip target a total of 10 times, interspersed with short rest periods. At the end of the sending period, the experimenter—who, like Jill, is blind to the target—takes Jill out of the ganzfeld condition and then discusses her impressions with her while they both look at four possible videos, one of which was the real target, and three were decoys. Jill is asked to rank the four videos based on her impressions of the target. By chance she'll rank the correct target first one in four times, for a 25% chance expected hit rate.

Psychologists Charles Honorton, William Braud, and Adrian Parker independently developed this technique in the 1970s. These ganzfeld tests have generated more debate and scrutiny among scientists than any other class of modern psi experiment.[22] One consequence of all this attention is that the modern ganzfeld experiment is as close to the perfect psi experiment as anyone knows how to conduct. Until recently, the ganzfeld experiments were largely unknown outside of the discipline of parapsychology. Then, in 1994, psychologists Daryl Bem from Cornell University and Charles Honorton from the University of Edinburgh published a meta-analysis of ganzfeld studies in *Psychological Bulletin*, a well-regarded academic psychology journal.[23] That paper provided strong evidence for a genuine psi effect. Bem and Honorton's review of earlier ganzfeld studies estimated an effect with overall odds against chance of 48 billion to 1, and their review of newer, fully automated experiments that were specifically designed to overcome all known criticisms of the previous studies, was also significant, with odds against chance of 517 to 1.

At the end of their report, Bem and Honorton concluded that new experiments conducted by other investigators would eventually resolve the question as to whether genuine psi was being observed in the ganzfeld experiment. Other scientists continued to conduct their own versions of these experiments, some similar to the "classic" ganzfeld tests, others using a variety of new procedures. Then in 1999, psychologists Julie Milton from the University of Edinburgh and Richard Wiseman from the University of Hertfordshire, England, published a new meta-analysis, again in *Psychological Bulletin.* In reviewing 30 new ganzfeld studies published after the Bem and Honorton paper,[24] they found a positive result, but it was so close to chance that they concluded that psi wasn't repeatable after all. Upon reading their analysis, I noticed that they had used a statistical method that underestimated the overall effect. In fact, the studies they selected were not just positive, but in statistical terms, they were *significantly* positive.[25]

A few years later, psychologists Lance Storm of the University of Adelaide, Australia, and Suitbert Ertel of Georg-August-University of Göttingen, Germany, formally answered Milton and Wiseman's paper.[26] They found that Milton and Wiseman had overlooked a number of earlier ganzfeld studies, and they argued that the best way to judge if the ganzfeld method really was successful was to combine *all* known studies. They found 79 studies that Bem and Honorton hadn't considered. This new batch of studies was associated with overall odds against chance of 131 million to 1.[27] One might think this would have settled the issue, but of course the debate vigorously continued, back and forth, like playing a game of badminton with hand grenades.[28] A portion of the debate hinged on whether one large-scale, highly significant study by University of Edinburgh psychologist Kathy Dalton was considered.[29] If it was, then everyone agreed that overall the evidence for psi in the ganzfeld was significant.

Then a new twist was added to the debate. Psychologist Daryl Bem and his colleagues noticed that there were two basic types of ganzfeld studies—those based on "standard" designs, like the studies conducted in the 1980s using picture targets, and newer experiments using "nonstandard" designs, like those using musical targets.[30] The latest round of debates over the ganzfeld tended to focus on nonstandard designs, and those studies had resulted in poorer performance. This might be because psi doesn't exist, or because the ganzfeld procedure cannot produce psi repeatedly, or—and this was the key—maybe people "get" certain kinds of psi information more easily than others. To test this idea, Bem examined the batch of more recent ganzfeld studies and found that the standard experiments using visual targets were significantly over chance with odds of 5,000 to 1, but the nonstandard studies using musical targets and other new variations were at chance.[31]

I've skipped over portions of this debate that concentrated on potential design flaws, as these arguments can be mind-numbingly tedious and they've often focused on trifling differences in procedures that haven't been shown to actually make any difference. This isn't to say that flaw analyses are a complete waste of time. Historically, identification of real flaws has been valuable in helping to refine experimental designs. But in recent years, the concept of *potential* flaw, regardless of how exceedingly unlikely it may be, has become a convenient "out" for those who prefer not to accept the evidence on any terms.[32]

GANZFELD META-ANALYSIS

From 1974 through 2004 a total of 88 ganzfeld experiments reporting 1,008 hits in 3,145 trials were conducted.[33] The combined hit rate was 32% as compared to the chance-expected 25% (Figure 6–6). This 7% above-chance effect is associated with odds against chance of 29,000,000,000,000,000,000,000 (or 29 quintillion) to 1. The funnel plot for the ganzfeld trials shows that selective reporting is not a problem for these studies (Figure 6–7).

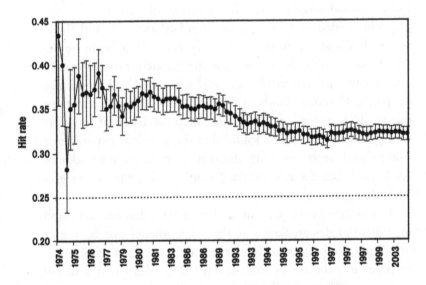

Figure 6–6. Cumulative average hit rate in the ganzfeld experiments, from 1974 through 2004, with one standard error bar. The overall odds against chance for these studies are 29 quintillion to 1, so chance is not a viable explanation for these results. Each dot represents an experiment and the dates on the x-axis indicate the average year of study publication.

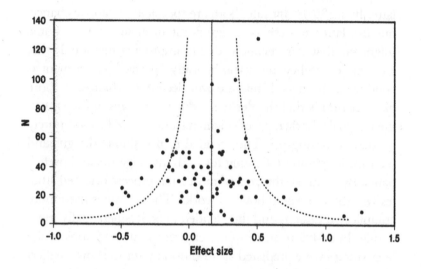

Figure 6–7. Funnel plot for the ganzfeld studies. The symmetric shape shows there's no file drawer problem.[34]

If we insisted that there *had* to be a selective reporting problem, even though there's no evidence of one, then a conservative estimate of the number of studies needed to nullify the observed results is 2,002.[35] That's a ratio of 23 file drawer studies to each known study, which means that each of the 30 known investigators would have had to conduct but not report 67 additional studies. Because the average ganzfeld study had 36 trials, these 2,002 "missing" studies would have required 72,072 additional sessions (36 x 2002). To generate this many sessions would mean continually running ganzfeld sessions 24 hours a day, 7 days a week, for 36 years, and for not a single one of those sessions to see the light of day. That's not plausible.

DECLINE EFFECTS

When we compare the first 44 ganzfeld experiments (average year of publication was 1981) with the last 44 studies (average year 1998), we find that the former had a 34.4% hit rate and the latter had a 30.3% hit rate. Both are spectacularly above chance, but the latter represents a significant drop in hit rate. Some might say that this decline is due to improving methodologies, so some day when the hypothetically "perfect" experiment is conducted the overall hit rate will decline to chance. A more likely reason is that the decline is due to changes in the experimental goals. Earlier, proof-oriented studies were focused purely on demonstrating psi. They tended to use simple designs and were exciting and motivating for both investigators and participants. By contrast, the goal of most later, process-oriented studies was to understand how psi works. Those studies used more complicated designs and they tended to be less personally motivating. In addition, some of the experimental conditions in the later studies were predicted *not* to show any (or to show smaller) psi effects. And thus when those studies were wrapped into the grand meta-analysis, the overall results would be expected to decline.

Why are decline effects interesting? Because a frequent observation in psi research is that when a new experiment is first conducted the outcomes are strikingly successful. Then, as others try to replicate the effects they begin to fade. Sometimes even the original investigators start to have problems replicating their own work. Are such declines unique to psi research, or do they occur in other experimental domains? The answer is important because if declines only occurred in psi research, it would raise suspicions that something's uniquely fishy about these experiments.

The evidence indicates that psi research is not unique. Meta-analyses in other disciplines also show declines. For example, an article published in the *Proceedings of the Royal Society* showed that effects reported in meta-analyses in the biological sciences often declined over time.[36] The article reviewed 44 meta-analyses published in peer-reviewed ecological and evolutionary biology journals. Effects across the 44 meta-analyses dropped significantly, with odds against chance of 250 to 1.

As another example, consider the results of experiments measuring milk production in dairy cows using a treatment designed to eliminate parasites (Figure 6–8). In 75 studies published from 1972 through 2001, there was a highly significant decline in the benefit of this treatment. One reason for the decline is that we're not dealing with a highly stable object like a rock, but with a complex system that is exquisitely sensitive to interactions among cows, parasites, treatment, and environment. The same is undoubtedly true for psychological processes. Psi in particular is the poster child for a highly dynamic and interactive process, thus it would be *surprising* if psi effects remained rock-steady over time.

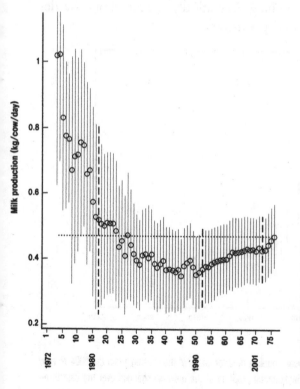

Figure 6–8. Decline in average milk production in dairy cows in response to an antiparasite treatment, from 1972 to 2001, with 95% confidence intervals.[37] Similar declines are frequently observed in psi experiments.

By comparison, one might expect that carefully measured properties of elementary physical particles, like the neutron, are much more stable. But are they? The "weak coupling ratio" associated with the decay of the neutron significantly dropped from the first measurements taken in 1969 to the most recent in 2001 (Figure 6–9).[38] This either means that a fundamental property of the universe has changed over the past three decades, or that measurement techniques have improved. The latter is probably a better explanation, but the difference between the first and last measurements differs by over 10 standard errors. That's a stupendously large change, many times larger than the decline observed in the ganzfeld psi studies. That said, I don't want to push this issue too far—the key point is that declines in measured effects have been observed across many scientific disciplines, so there's nothing dramatically unique about the declines observed in psi experiments.

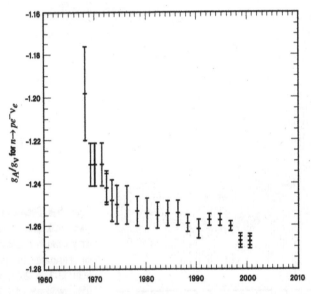

Figure 6–9. Decline in the weak coupling ratio for decay of the neutron. From the 2004 *Review of Particle Physics*. Has a fundamental property of the universe changed over the past three decades, or have measurement techniques improved?

So far we've seen that the dream and ganzfeld experiments each provide strong evidence for the existence of psi. Together they represent over 4,000 sessions conducted under controlled conditions, by dozens of investigators worldwide, over many decades. The experimental designs have been repeatedly critiqued and refined to achieve the most rigorous standards, and there are no discernable problems of selective reporting. These two databases should be sufficient to convince anyone that psi exists. But there's more. Much more.

THE SENSE OF BEING STARED AT

Terms like *telepathy* and *clairvoyance* tend to imply that psi consists of a few tidy, well-defined abilities. But the implication is misleading. Psi refers to a general process of information transfer, and telepathy and clairvoyance are merely two of the innumerable ways that this phenomenon can manifest. Another way is the sense of being stared at, a lore associated with the power of the gaze. In its negative connotation, the sense of being stared at is associated with the "evil eye," one of the oldest and still-prevalent superstitions in modern times. The evil eye refers to the belief that too much attention paid to an object or an individual would spark desire, and that intention would in turn lead to envy, jealousy, bewitchment, and in general, evil. The word "fascination" is closely related to this belief. Its etymology can be traced to the Greek *phaesi kaino,* meaning "to kill with the eyes."

In 1895, the British folklorist Frederick Thomas Elworthy published a classic work on this topic, entitled appropriately enough, *The Evil Eye*. He described as universal the belief that the eye has the power to emit an emanation or force, and that a malignant influence may dart "from the eyes of envious or angry persons, and so [infect] the air so as to penetrate and corrupt the bodies of both living creatures and inanimate objects." Elworthy's book was written over a century ago, and yet we find that his opinion still rings true today. As he put it (in 1895):

> We in these latter days of Science, when scoffing at superstition is both a fashion and a passion, nevertheless show by actions and words that in our innermost soul there lurks a something, a feeling, a superstition if you will, which all our culture, all our boasted superiority to vulgar beliefs, cannot stifle, and which may well be held to be a kind of hereditary instinct.

All you need to do to confirm Elworthy's statement is search the Internet for "evil eye amulets." You'll quickly locate tens of thousands of web pages discussing or selling trinkets, bracelets, and charms to ward off the evil eye.[39] From a scientific perspective, the question is—as it is throughout this book—do such widespread beliefs have any basis in fact? Or is it all merely superstition borne by ignorance or anxiety?

The sense of being stared at has been experimentally investigated for nearly a century. In the typical study, one person does the staring, let's call him Jack, and another is stared at, let's call her Jill. Jack and Jill sit within a few yards of each other, Jill with her back to Jack. Jack flips a coin to determine on each successive trial whether he should stare or not stare at the back of Jill's head.[40] If the assignment is to stare, then for 10 seconds Jack intensely stares at Jill. Then he alerts Jill with a clicking tone to respond "yes" if she thinks Jack has been staring at her, or "no," if she thinks he hasn't.

British biologist Rupert Sheldrake has popularized experiments based on this simple design, some involving trial-by-trial feedback under casual conditions, such as tests conducted by pairs of children in classrooms, and under more controlled conditions, such as those involving use of blindfolds, no trial-by-trial feedback, and even more secure conditions such as having Jack stare at Jill through a window from a distance.[41] I found 60 such experiments involving a total of 33,357 trials from publications cited by Sheldrake and others.[42]

The overall success rate in these experiments was 54.5% where chance expectation is 50% (Figure 6–10). The overall odds against chance are a staggering 202 octodecillion (that's 2 x 10^{59}) to 1.

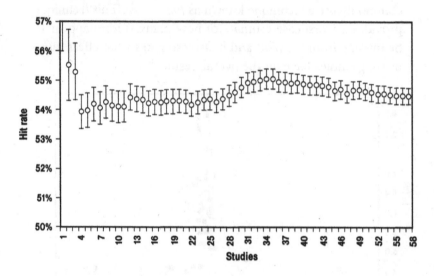

Figure 6–10. Cumulative average hit rate and one standard error bar in 60 sense-of-being-stared-at experiments involving a total of 33,357 trials. The odds against chance (at 50%) are 202 octodecillion to 1, definitively ruling out coincidence as an explanation.

Many critiques have been offered as mundane explanations for these results. They involve the usual array of suggested flaws, fraud, and selective reporting.[43] Many of the suggested flaws have been tested and found not to provide plausible explanations. Some flaws can explain, at least in principle, *some* of the results. But no individual flaw or combination of flaws has been found that can credibly account for the overall results. Likewise, collusion as an explanation is implausible because many independent groups have successfully replicated these results.

Selective reporting is a more serious possibility, as many of these studies were conducted by groups in schools, and it's likely that some of those studies weren't reported. A funnel plot analysis of these studies indicates that there *is* indeed a selective reporting problem. There are too few small, negative effects in the funnel plot (Figure 6–11). To automatically identify and fill in the missing "bite" taken out of a funnel plot, statisticians commonly use a technique known as *trim and fill*. This technique provides a worst-case estimate of how many studies appear to be missing from the plot, and it also estimates what effect those missing studies have on the overall results.[44]

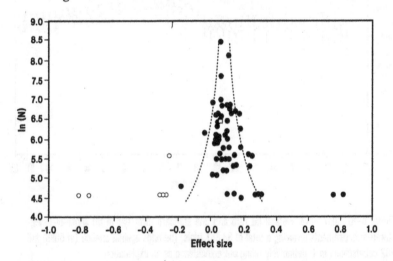

6-11. Funnel plot for the sense-of-being-stared-at experiments (black circles). A replication study conducted at the Institute of Noetic Sciences is shown as the small white box in the center of the funnel. The trim and fill analysis estimates that six studies are probably missing (white circles).[45]

The trim and fill technique estimates that six studies are probably hidden away in file drawers. When those studies are added to the mix, we find that the overall results are reduced slightly.[46] But the odds against chance are still astronomically beyond coincidence, at 10^{46} to 1.

To completely nullify this outcome, we'd need 1,417 non-significant studies to be hidden away in file drawers. This is most implausible, so selective reporting cannot explain the results. Analysis of differences in the controls employed in these experiments indicate that studies that didn't rigorously shield against sensory cues ended up with larger effects than studies with better controls. This means some portion of these overall results *might* be due to subliminal sensory cues picked up by Jill.[47] But a subset of the highest-security studies still provides abundantly significant outcomes, and so some of the sense-of-being-stared-at experiments do appear to involve psi.

The class of studies we've considered so far are based on experimental designs that, in the final analysis, ask a participant to select one randomly chosen target out of a pool of possible targets. This hit vs. miss approach is used because it simplifies the question of whether psi exists by turning the answer into the equivalent of a coin flip. These studies show us that we're getting far more hits than expected by chance. But the coin-flip approach also compresses the richly detailed information generated in the dream and ganzfeld studies into a single point, and it creates the illusion that psi-type information is weak and highly variable. Many researchers today are beginning to look more closely at the actual content of the dreams and ganzfeld mentation to learn *why* and *when* certain information transfers from Jack to Jill. They're studying special populations like highly creative artists and musicians, who provide much larger effects in these experiments. And new variations of the sense-of-being-stared-at experiment, like the sense of anticipating another person's arm movement, are being explored.[48] These studies suggest that our ongoing, ordinary stream of conscious awareness may be masking a much larger repository of psi residing in our *unconscious*. Let's see if the evidence supports this idea.

CHAPTER 7

UNCONSCiOUS PSi

In studying the history of the human mind one is impressed again and again by the fact that the growth of the mind is the widening of the range of consciousness.

−Carl Jung

Studies in psychotherapy and in cognitive neuroscience indicate that conscious awareness is like a stream of water trickling through a crack in an immense dam. If psi appears in the tiny trickle of conscious experience, then what resides in the depths of the unconscious? Could this be where intuitive hunches, gut feelings, and premonitions come from? To find out, we'll look at classes of experiments studying unconscious psi effects in the human body. We'll focus on studies involving three aspects of the nervous system—the part that regulates automatic functions like heart beat and sweating (called the autonomic nervous system), the part involved in conscious movements and thought (the central nervous system), and the part involved in digestion and elimination (called the enteric nervous system).

PSI IN THE AUTONOMIC NERVOUS SYSTEM

This first class of studies is known by the acronym DMILS, which stands for "direct mental interactions with living systems." In a DMILS study, when Jack and Jill arrive at the laboratory, the experimenter escorts Jill to a room that looks like a large walk-in freezer. This solid steel, double-walled chamber shields against electromagnetic signals and acoustic noise. Such rooms are used to ensure that no ordinary forces or signals can reach Jill after the door to the chamber is closed.[1] In contrast to the intimidating, cold steel exterior of the chamber, the interior of these rooms is often decorated in earth tones and silk plants to make it look warm and inviting. Jill is asked to sit in a comfortable reclining chair in the chamber, where the experimenter wires her up to a monitor that measures changes in the activity of her sweat glands.[2] This activity is regulated by the autonomic nervous system and is a convenient way to measure changes in Jill's emotional state.

Once wired up, Jill is asked to simply relax for about 30 minutes while she is continuously monitored.[3] Her only task is not to fall asleep, and to think about Jack to try to maintain a mental connection with him. Jill knows that Jack will be thinking about her from a distant location, but she doesn't know when, or how long, or the type of thoughts he may direct towards her. After the experimenter confirms that Jill's skin conductance data is being recorded properly, the chamber door is closed. These chambers are designed to create a tight electrical seal, so shutting the door is like sealing the hatch on a spacecraft. (Note to student investigators: It's useful to inform Jill that after this massive door is closed she can still get out.)

Now Jack is escorted to a distant, soundproofed location and asked to sit in front of a video monitor and follow the instructions that pop up on the screen. The computer in another room that is controlling the experiment waits a few minutes, then it decides, based on the equivalent of a coin toss, if it should instruct Jack to *calm* Jill, or to *activate* her. If it decides that Jack should calm Jill, the word "calm" pops up on the monitor, instructing Jack to imagine Jill in a relaxing, calming setting, like taking a nap on a beach. If he was trying to activate her, he might imagine that she was running up a steep hill, or skydiving. When the pop-up instruction disappears, typically in 20 seconds or so, Jack withdraws his attention from Jill and the computer starts a timer to wait for the next trial. Over the course of 30 minutes, the computer might be programmed to present a total of 20 calm and 20 activate instructions, in a random order. In some experimental setups, Jack can watch a strip-chart record of Jill's ongoing skin conductance changes that he can use as feedback to adjust his mental "sending strategy" to influence Jill.

At the end of the experimental session, the investigator takes the continuous, 30-minute record of Jill's skin conductance data and separates it into two subsets: those periods when Jack was aiming calming vs. activating thoughts towards Jill. If it turned out that when Jack was thinking calming thoughts Jill showed lower skin conductance activity, and vice versa for activating thoughts, and this relationship persisted over the course of many test sessions, then the experiment would have demonstrated an unconscious psi connection between Jack and Jill. To fully justify this inference, care must be taken to ensure that no ordinary signals can pass between the pair, and that Jill doesn't know when, or in what way, Jack is trying to influence her. Properly conducted experiments employ these and many other controls.[4]

A variation of the DMILS experiment is known as a study in "remote staring." This is a rigorously controlled version of the sense-of-being-stared-at experiment described in the last chapter.[5] At randomly selected times, Jack sees Jill's live video image over a closed-circuit TV monitor (Figure 7–1). When this happens he stares intently at Jill, aiming to activate her nervous system. When the screen goes blank, he relaxes and thinks about something else. Unlike the conscious sense-of-being-stared-at-experiment, Jill is not asked to report what she thinks is happening. Instead, unconscious fluctuations in her skin conductance are used to judge whether she's reacting to Jack's distant stare.[6]

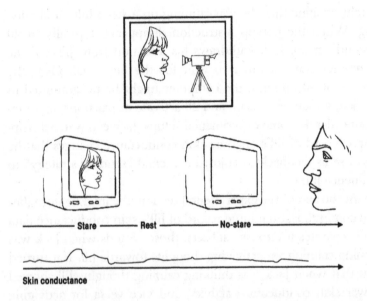

Figure 7-1. Distant-staring experiment design. Jill relaxes in front of a video camera for about a half hour. When her image appears at random times on a video monitor in front of Jack, he tries to mentally connect with her and activate her nervous system. Changes in Jill's skin conductance when she is being stared at vs. not being stared at are used to test whether Jack's intention influences Jill.

DMILS META-ANALYSIS

In 2004 psychologist Stefan Schmidt and his colleagues from the University of Freiburg Hospital, Germany, published a meta-analysis of these two classes of experiments in the *British Journal of Psychology*.[7] They found 40 DMILS studies reporting 1,055 individual sessions conducted between 1977 and 2000. The overall results were significant with odds against chance of 1,000 to 1, so coincidence is not a viable explanation.[8] The funnel plot found no selective reporting bias (Figure 7–2). And no significant relationship was found between experimental quality and the resulting outcomes, so the results were not due to flaws in the experiment.

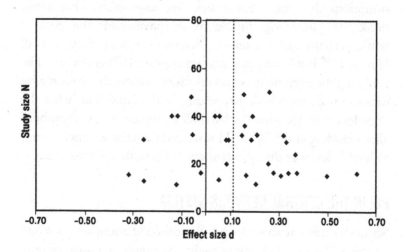

Figure 7–2. Funnel plot for DMILS studies (Schmidt 2004). The dashed vertical line shows the average effect size, the solid line is at chance. There is no evidence of selective reporting.

For the set of remote-staring studies, Schmidt's team found 15 experiments describing a total of 379 sessions conducted between 1989 and 1998. As in the DMILS studies, the meta-analysis found a significant effect with odds against chance of 100 to 1,[9] no selective reporting, and no relationship between study quality and outcome. In discussing their findings, Schmidt's group noted that "because of the unconventional claim of the studies under research, we always chose a more conservative strategy whenever such a decision had to be made." They concluded that for both classes of experiments, "There is a small, but significant effect. This result corresponds to the recent findings of studies on distant healing and the 'feeling of being stared at.' Therefore, the existence of some anomaly related to distant intentions cannot be ruled out."[10]

This last sentence is written with the proper cool, scientific tone. But it's an astounding conclusion appearing in an academic psychological journal, especially in light of centuries of assumptions that such "anomalies" are impossible. One might think this interesting finding would have made the evening news, perhaps with the teaser, "Scientists prove evil eye is real! Film at 11!" But hardly any reports appeared. This is a little like watching the evening news on television, where the news reader drones on about what's happening in the latest war, what the President is up to, what the baseball scores are, something about aliens landing at the White House, and then the weather report. What? What was that about aliens? Oh, nothing important.

PSI IN THE CENTRAL NERVOUS SYSTEM

Schmidt's meta-analysis found that thinking about another person at a distance influences their autonomic nervous system. Does thinking about distant people also cause changes in their brains? Given the evidence for telepathy, one might predict that the answer must be yes. What does the experimental evidence say?

The design used in these electroencephalograph or "EEG correlation" experiments asks, in effect, whether poking one person will produce an *ouch* response in a distant partner. It's not recommended to poke people in the brain, so instead we use a stimulus like a flashing light to cause one of the brains to jump electrically in a predictable way, and then we look at the other, distant brain to see if it's jumping at the same time. I'll discuss one of these experiments in more detail later to explain how it works, but let's first briefly review the history of these studies.

The first two experiments investigating EEG correlations in separated pairs of people were reported in the 1960s. The first study was conducted by psychologist and "altered states of consciousness" pioneer Charles Tart at the University of California, Davis. The second involved identical twins and was published in the prominent scientific journal, *Science*.[11] Those two articles soon generated a flurry of ten conceptual replications by eight different groups around the world. Of the ten studies, eight were reportedly positive.[12] One of those studies was published in the top-ranked journal, *Nature,* and another appeared in the mainstream journal, *Behavioral Neuroscience.*

A decade later, psychophysiologist Jacobo Grinberg-Zylberbaum and his colleagues from the National Autonomous University of Mexico reported a series of studies in which they claimed to detect simultaneous brain responses in the EEGs of separated pairs of people.[13] One of their studies was published in the journal *Physics Essays,* stimulating another round of replication attempts.[14] In 2003, a successful replication was reported in *Neuroscience Letters* by EEG specialist Jiří Wackermann and his colleagues. They attempted to close all known loopholes in the previous studies and applied a sophisticated analytical method to the resulting brain-wave data. Wackermann's team concluded that

> We are facing a phenomenon which is neither easy to dismiss as a methodological failure or a technical artefact nor understood as to its nature. No biophysical mechanism is presently known that could be responsible for the observed correlations between EEGs of two separated subjects.[15]

Another successful replication, this time reported by Leanna Standish of Bastyr University and her colleagues, was recently reported in the medical journal, *Alternative Therapies in Health and Medicine.* They conducted an EEG correlation experiment with the receiving participant located in a functional magnetic resonance imaging (fMRI) scanner.[17] They prescreened 30 pairs of people to find one couple that was able to reliably produce a correlation, then they put one person acting as the receiver in an fMRI scanner and the other in a distant room. They found a highly significant increase in brain activity (odds against chance of 14,000 to 1) in the receiving person's visual cortex (in the back of the brain) while the distant partner was viewing a flickering light (Figure 7–3). The same group later successfully replicated this finding.[16]

This means not only that a significant correlation was observed between two brains, but also that the precise location in the brain associated with this connection was found. This discovery is so shocking that it virtually guaranteed no one would hear about it, despite it being published in a medical journal. This is worse than missing a story about aliens landing on the White House lawn. It's more like spotting an alien shopping in the frozen food section of the local grocery store and no one caring.

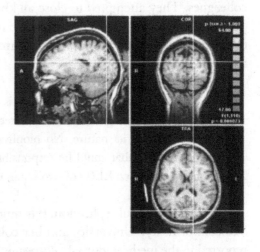

Figure 7-3. The crosshairs show where the "receiver's" brain was significantly more active while a distant "sender" was viewing a flickering checkerboard screen. This activation, appearing in the receiver's visual cortex, suggests that the receiver's brain was mimicking the sender's brain, as the sender's visual cortex would have been considerably activated by the flickering stimulus.

There's more. In 2004, three new independent replications were reported. All were successful. One appeared in the *Journal of Alternative and Complementary Medicine,* by Leanna Standish and her colleagues. She tested 30 pairs of people who were trained in a meditation technique, and found overall odds against chance, in favor of real EEG correlations, at 2,000 to 1.[17]

University of Edinburgh psychologists Marios Kittenis, Peter Caryl, and Paul Stevens reported a second experiment.[18]

They tested 41 volunteers, of which 26 were emotionally bonded couples, 10 were randomly matched strangers, and 5 were individuals who *thought* they were being paired up with someone they hadn't met yet, but in fact they were run in the experiment alone. Kittenis's team found a significant increase in the magnitude of the EEG alpha rhythm for the related pairs (odds of 50 to 1) and also for unrelated pairs (with odds of 143 to 1), but not for the 5 participants with no distant partner. Kittenis's team also found, by comparing brain-maps showing patterns of electrical stimulation in the sender and receiver, that the receivers' brains showed patterns of activation that closely mimicked the activation in the senders (Figure 7–4).

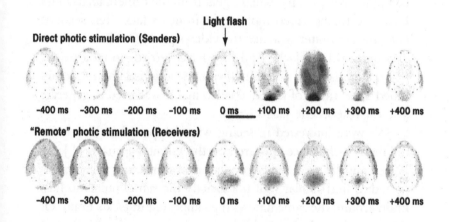

Figure 7–4. Alpha rhythm power in the average sender's brain (top) and average receiver's brain (bottom). The stimulus, a light flash, appears at time "0 ms" (milliseconds). The dark spots appearing in the receivers' brain are patterns of activation mimicking the much larger responses in the senders' brain. This shows an unconscious, extrasensory connection between two brains.

The third EEG correlation study was one my colleagues and I conducted at the Institute of Noetic Sciences (IONS).[19] We recruited 13 pairs of friends for this study. We didn't require any special relationships, just a shared interest in participating in the experiment. When the pair arrived at the laboratory, they mutually decided who wanted to be the sender (Jack) and the receiver (Jill). After attaching EEG electrodes to Jack and Jill, we asked Jill to sit in a reclining chair inside our electromagnetically and acoustically shielded chamber, and then we escorted Jack to a dimly lit room about 30 feet away, behind three doors.[20] A closed-circuit TV camera was focused on Jill's face in the shielded room.

With Jack and Jill safely ensconced in their separate rooms, and their electrodes wired up to separate EEG amplifiers, I started a computer program that ran the rest of the experiment automatically. At the beginning of each "sending" period, the computer switched the video signal from the camera focused on Jill's face to the video monitor in front of Jack. Ten seconds later, the computer switched the video signal off. The computer also marked both Jack's and Jill's EEG records to indicate the beginning and end of these 10-second periods. The randomly timed appearance of Jill's live video image was used to generate a startle response in Jack's brain.

We were interested in seeing what happened in Jill's brain the moment her face appeared on the monitor in front of Jack. If Jack and Jill were really connected in some way, then we expected to find similar (but not exactly the same) pulses in Jill's brain around the same time as they appeared in Jack's brain. Of course, neither Jack nor Jill knew in advance the number of sending periods, when they would occur, or even the precise length of the experimental session.

The results confirmed that Jack's brain jumped in response to the sudden video stimulus (not his actual brain, but rather the electrical activity of his brain), as we expected. It took about a third of a second for the brain-wave activity to reach its peak, which is the time-length expected based on many previous neuroscience studies in "visual evoked potentials."[21] In addition, as predicted by a psi connection, we saw that Jill's EEG peaked within milliseconds of Jack's EEG peak. The correlation between Jack's and Jill's brain responses was positive, with odds against chance of 5,000 to 1 (Figure 7–5).[22] To make sure that this result wasn't due to an equipment or analytical problem, we conducted the same experiment to see if the electromagnetic pulse caused by the video screen suddenly turning on in Jack's room was picked up by the sensitive EEG amplifier in Jill's shielded room. The pulse did not appear, so the original correlation reflected a genuine relationship.[23]

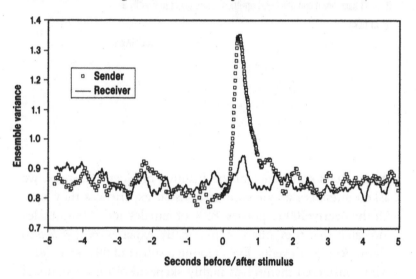

Figure 7–5. Brain-wave responses for sender and receiver EEGs averaged across 13 pairs of people; in each case the sender in the pair was exposed to 25 visual stimuli. On average, the senders' EEG peaked 392 msec after the presentation of the stimulus, and the receivers' EEG peaked 64 msec later. The odds against chance of this relationship is 5,000 to 1.[24]

So psi can be detected unconsciously in both the autonomic and central nervous systems. What about the nervous system that controls the gut? Is there a psi connection with *gut feelings*?

CHAPTER 8
GUT FEELiNGS

When you feel in your gut what you are and then dynamically pursue it—don't back down and don't give up—then you're going to mystify a lot of folks.

—Bob Dylan

At the 2003 conference of the Institute of Noetic Sciences, we asked attendees about various unusual experiences they had. Of the nearly 500 responses, 89% of females and 72% of males indicated that they often or frequently experienced *gut feelings* about people or events. Even among a subset of 89 respondents who considered themselves highly skeptical of unconventional claims, 78% reported that they also often experienced gut feelings.[1] Sometimes gut feelings reflect no more than a bad burrito or emotional turbulence. But could some gut feelings, which even skeptics admit to, also include psi information?

Intuitive hunches and psi experiences all involve the act of knowing without knowing how you know, and gut feelings in particular imply a form of intuition based on visceral sensations in the belly. From a conventional point of view, intuitive hunches and gut feelings are due to factors such as forgotten expertise, subliminal cues, and unconscious inferences.[2] However, if intuition is related to psi, then it's possible that some gut feelings might also carry psi information.[3] To test this idea in the laboratory, we ran an experiment similar to the studies discussed in the previous chapter, except instead of using an EEG we used an EGG, an electrogastrogram.[4] The EGG measures the electrophysiology of the gut, which is a slow rhythm of about 3 cycles per minute.[5]

The gut is a particularly interesting portion of the nervous system to study because of its close relationship with emotions. Phrases like "butterflies in the stomach," "a gut-wrenching experience," and "a sinking feeling in the stomach," attest to this familiar correspondence, which has been studied for nearly two centuries.[6] We wondered if gut feelings might be especially sensitive to detecting *emotions*-at-a-distance.

In this study, the sender—Jack—sat in front of *two* video monitors and wore a set of headphones. At random times, one video monitor displayed the receiver's—Jill's—live video image for two minutes while the other showed a sequence of emotional or neutral pictures as emotionally appropriate music played over the headphones. When Jill's image disappeared, both monitors faded to black and the music stopped. Between each emotional condition there was a 30-second rest period.

The pictures used to evoke positive emotions in Jack included colored photos of smiling babies, kittens, and appetizing food. When those pictures appeared, they were accompanied by an upbeat song like "Twist and Shout" by the Beatles. There were two types of negative emotions evoked: angry and sad. The angry condition consisted of colored pictures like an atomic bomb explosion, accompanied by the angry-sounding song, "Feuer Frei," by the heavy metal rock band, Rammstein. The sad condition consisted of pictures such as a graveyard and unhappy people, accompanied by Samuel Barber's "Adagio for Strings." A calm condition used black and white pictures like a plain soup bowl, accompanied by the song, "May It Be," by Enya. The emotionally neutral condition consisted of a series of gray rectangles with minor differences in hues, accompanied by pink noise.

Jack was instructed to periodically gaze at Jill's image while trying to mentally "send" the emotions evoked by the slide show and music. Between sending periods, Jack was instructed to withdraw his attention from Jill and just relax. We expected that if gut feelings involved a type of psi perception, then we'd find that Jill's gut was more active during the emotional conditions (her stomach would get "tied up in knots") as compared to the neutral conditions.

We ran 26 adult couples through this experiment. All of the couples knew each other, some casually as friends and others as long-term partners; each pair mutually decided who would play the role of Jack and Jill. The results showed that Jill's EGG responses were significantly larger when Jack was experiencing positive and sad emotions than when he was experiencing neutral emotions (with odds against chance of 167 to 1, and 1,100 to 1, respectively).[7] Most of the churning in Jill's gut occurred within 20 seconds of the beginning of the emotional period (Figure 8–1).

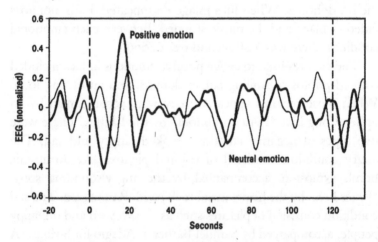

Figure 8–1. Average gut responses, as measured by an electrogastrogram, for all positive (bold line with the large response at 20 seconds) and neutral emotional conditions (thin line), across all 26 receivers in the experiment. This shows that the receiver's gut clenched more while the distant sender was experiencing positive emotions than when experiencing neutral emotions.

We considered many alternative conventional explanations for these findings. Leading candidates included chance, a violation of statistical assumptions, sensory cues, expectation biases, and physiological drifts that might have coincidentally matched the emotional conditions. Each explanation was evaluated and rejected as inadequate.

This experiment suggests that sometimes gut feelings respond to the emotional states of distant people. This in turn implies that some decisions influenced by visceral and other somatic feelings may involve psi perception. It would be rash to assume that *all* gut feelings are infused with intuitive information, as many things can lead to the odd visceral twitch. But it may turn out that the belly brain's intuition is more connected with the rest of the world, and with other people, than previously suspected.

So far, the studies we've reviewed indicate that when Jack mentally interacts with Jill at a distance she can perceive that information both consciously and unconsciously. But what these studies don't tell us is *how* this connection works. Does Jill passively perceive Jack's intention and respond accordingly? Or is she literally *influenced* by Jack in some way? To see if the latter is possible in principle, let's investigate the evidence for direct interactions between mind and matter.

CHAPTER 9
MiND-MATTER iNTERACTiON

The universe begins to look more like a great thought than like a
great machine.

—Sir James Jeans

We've been considering the question of psi as a perceptual ability, as information flowing from the environment into the mind
without the use of the ordinary senses. This suggests a relationship that might be written as matter→mind. What about information flowing in the other direction? Could this relationship
be symmetric? Could there also be a mind→matter connection,
a psi *influence*?

At first glance, information and influence appear to be quite
different. The former is passive and involves subjective issues
like knowing and understanding; the latter is active and involves objective concepts like force and energy. However, as
physics has evolved, apparent differences between information
and influence have also dissolved. Today, some physicists are
entertaining the possibility that reality might be literally constructed out of information.[1] The eminent physicist John
Archibald Wheeler coined the pithy phrase, "It from bit," to
refer to the quantum perspective of how the universe appears to
be made out of bits of information rather than bits of matter or

energy. Wheeler proposed that we live in a participatory universe in which we—our act of asking questions of Nature—participate in the creation of the observed world. As he put it, referring to physics experiments,

> " . . . every it—every particle, every field of force, even the spacetime continuum itself—derives its function, its meaning, its very existence entirely—even if in some contexts indirectly—from the . . . answers to yes-or-no questions, binary choices, *bits*." [2]

PSI IN DICE

In 49 BC, Julius Caesar and his army crossed the river Rubicon to invade Italy. He reportedly shouted to his troops while crossing the river, "The die is cast!" By this he meant that his decision to invade Italy had rolled a die, and history would record their fate. The toss of the die was favorable to Caesar that day, as his decision ultimately founded the Roman Empire, which in turn spawned Western civilization.

Caesar placed his fate in the hands of his army and the gods. Is it possible for us to mentally exercise control over how "the die is cast"? One might think the answer is no, for casinos enjoy healthy profits regardless of what gamblers are wishing. But the casino isn't a fair place to look; cash predictably pours into casinos because the odds are stacked in favor of the house. So if Julius Caesar could visit Caesar's Palace casino in Las Vegas, and he *was* able to mentally influence the dice to a small extent,

all he'd end up doing is losing the riches of the Roman Empire a bit slower than he might have if he couldn't influence the dice at all.

Nevertheless, many people act as though their

thoughts do affect the world. Surveys show that the vast majority of the world's population prays. A good percentage of those prayers ask Nature, or God, or the universe to cast the die favorably for oneself or for one's loved ones. The research discussed in previous chapters seems to support the idea that psi is a type of distant influence, in which case prayer could, in principle, affect the world directly. But the very same outcomes could also arise not because psi acts as a wish-modulated distant *influence,* but because the receiving party *perceives* intentions or actions of the distant person and responds accordingly.

So to test if distant influence is possible we need to see if nonliving objects, like dice, also respond to distant intention. Starting in 1935, researchers began to explore this idea. Over the next half century, 52 different investigators published 148 such dice-tossing experiments in English-language publications.[3] The term most often used for the postulated effect of mind over matter is *psychokinesis,* or PK for short.

The dice-tossing experiment is the epitome of simplicity. A die face is chosen in advance, then one or more dice are tossed while a person wishes for that face to turn up. If the person's intention matches the resulting die face, then a "hit" is scored. If more hits are obtained than expected by chance over many dice tosses, that's evidence for PK.

In spite of all the research and critical reviews of dice-tossing PK evidence collected over half a century, no clear consensus had emerged.[4] The controversy revolved around the belief that the PK effect is so difficult to repeat that *any* claim for PK must be considered suspect. Also, the apparently simple dice-tossing task obscures a bewildering array of potential pitfalls, any one of which could cast doubt on interpretations of the experimental results.

In 1989, when psychologist Diane Ferrari and I were at Princeton University, we used meta-analysis to assess the combined evidence for PK effects in dice experiments.[5] We searched through all the relevant English-language journals for dice experiments published from the 1930s to 1989. For each study we recorded the number of participants in the test, the die face they were aiming for, the number of dice tossed in one throw, and so on. We also coded each study for thirteen quality criteria, such as whether the study employed automatic recording, whether witnesses were present, and whether control tests were performed.

We found 73 relevant publications, representing the efforts of 52 investigators from 1935 to 1987. Over that half-century, some 2,500 people attempted to mentally influence 2.6 million dice throws in 148 different experiments, and just over 150,000 dice-throws in 31 control studies where no mental influence was applied to the dice. The total number of dice tossed per study ranged from 60 to 240,000, and the number of dice tossed in one throw ranged from 1 to 96.

While the overall effect was small in terms of absolute magnitude, it wasn't due to dumb luck.[6] The odds that the dice studies were due to chance alone were 10^{96} to 1 (that's 10 with 96 zeros after it). By contrast, the results of control experiments were well within chance expectation. So something else was clearly going on.

Maybe these results were due to just a few investigators who reported most of the studies, raising suspicions of fraud or sloppy work. To test this idea, we noticed that the number of studies conducted per investigator ranged from 1 to 21, with most investigators (64%) reporting three or fewer studies. So we calculated the overall odds against chance for that group of investigators. That subset still showed odds against chance of over a billion to one. So the results weren't due to a few suspect investigators. Maybe the results were due to a few experiments with impossibly good results? To test this possibility, we discarded 35% of the studies with the largest effects, and the remaining 96 studies still resulted in odds against chance of more than 3 million to one.

Perhaps the results were due to a selective reporting problem? To investigate this, I conducted a "trim and fill" analysis on the funnel plot, as discussed in previous chapters, and estimated there were 21 missing studies (Figure 9–1). When those studies were added to the funnel plot, the overall effect was adjusted downwards, but the odds against chance remained staggeringly high (10^{76} to 1).[7]

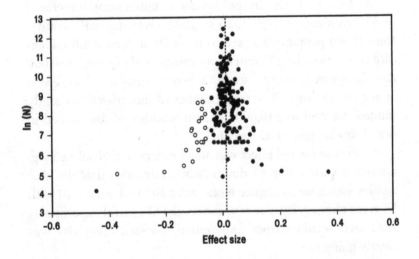

Figure 9–1. Funnel plot for 148 known dice experiments (black dots) and 21 missing studies estimated by the trim and fill algorithm (white dots). The dashed line shows the overall effect size. This tiny positive shift above chance is associated with odds against chance of 10^{76} to 1.

As a secondary check, we estimated the number of file-drawer studies necessary to bring the observed results down to chance. The number was 3,204, a ratio of 22 to 1 file-drawer to observed studies. That means that for selective reporting to explain these results we'd need each of the 52 different investigators to have conducted 62 studies, for all of those studies to have failed, and for none of them to have been published. This is implausible.

Maybe the results were due to systematically poor experimental quality? Psi research has always attracted a particularly passionate form of scrutiny, so these experiments, on average, tend to be more rigorously designed and executed than those in other fields.[8] In this particular case, one approach to assessing the effect of quality is to see whether the experimental designs improved over time. If these studies were conducted by crackpots, we wouldn't expect the quality to improve because crackpots don't pay any attention to critiques of their work. But an examination of the quality scores showed a clear positive trend over time, with odds against chance over a million to one. So we know that the researchers did take note of criticisms and progressively improved their work. With this knowledge in hand, if experimental results plummeted as experimental quality improved, then we would be justified in believing that the results might be due to flaws. We checked this by looking at the relationship between hit rates (averaged by year of publication) and the study quality averaged per year. The relationship was not significant, so quality variations aren't a plausible explanation either.[9] In sum, common explanations like chance, quality, and selective reporting cannot explain away these results.

IS IT INFLUENCE?

The evidence suggests that mind influences the fall of tossed dice. But what does influence mean in this context? Bouncing dice move so quickly that no one could keep track of each die as it was moving so as to mentally direct its behavior. Regardless of how we might imagine mentally "pushing" a die to fall as we wish, it seems exceedingly unlikely that the form of influence we're dealing with is anything like a simple, mind-directed *force.* Sorcerers in the movies are often portrayed as exercising PK by fiercely concentrating on an object, often accompanied by glowing "force beams" shooting out of their eyes like controlled lightning bolts. But nothing resembling that sort of influence has ever been observed in laboratory tests or outside the lab. So a different form of influence is necessary to explain mind-matter interactions. One alternative is that the act of observation might affect the probabilities of physical events occurring at the subatomic scale. This idea arises from interpretations of quantum mechanics, and we'll discuss it in more detail later. For now, imagine that the type of influence we're dealing with is not a mundane physical force, like maneuvering a die to force it to land with the desired number face up, but rather more like subtly altering the shape of the die, so the *likelihood* of the desired number landing face up is greater than that for the remaining five faces.[10]

We can test this idea by seeing what sort of effects we get when we toss one die, and then two dice, and then 20 dice, and so on. If the mind influences *each die* in approximately the same way, then as we toss more dice at once, the statistical yield *per toss* ought to increase.[11] Said another way, if our minds really can influence the fall of dice, then our ability to detect that PK effect ought to improve as we toss more dice at once.

In fact, when we analyze these studies according to how many dice were tossed at once (Figure 9–2), we find that the observed effects do indeed increase, and the relationship between the observed and predicted increase is significantly positive, with odds of 110 to 1.[12] If you disregard experiments involving 30 and 48 dice tossed at once (there were just three such experiments), the odds improve substantially, to 5,300 to 1.[13] This supports the idea that PK may be a type of influence.

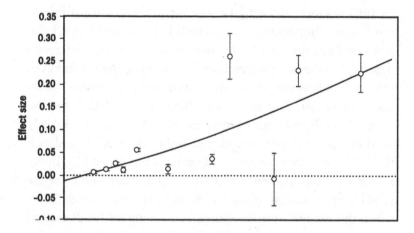

Figure 9–2. Effect size increases in the dice experiments as the number of dice thrown at once increases. The curved line is the best fit to the observed effects.[14] This suggests that mind directly influences matter.

All this should be heartening for gamblers who enjoy dice games because it suggests that what we wish is indeed reflected in the actual behavior of physical objects. It also confirms the faith of those who believe in the power of prayer. But hold on a minute. If all this is true, then why aren't the casinos going out of business? And why don't prayers work more reliably? The truth is that no one knows—yet. These experiments suggest that mind and matter are indeed related to a small degree that is statistically repeatable under controlled conditions. But we've barely scratched the surface of a phenomenon that's still profoundly mysterious. So offering answers to all the "but why" questions evoked by these data, given our present state of knowledge, is terribly premature. I think a more reasonable question to ask at this point is: If the results of the dice experiments suggest a genuine mind-matter interaction, then there ought to be corroborating evidence from similar experiments using other types of physical targets. And there is.

PSI IN RANDOM NUMBERS

In 1997, engineer Robert Jahn and his colleagues at the Princeton Engineering Anomalies Research Laboratory (PEAR Lab), published a review of 12 years of experiments in their lab investigating mind-matter interactions.[15] The experiments involved over 100 volunteers, all of whom attempted to mentally influence random number generators (RNGs). An RNG is an electronic coin-flipper that generates thousands of completely random coin-flips per second, but rather than heads and tails the RNGs generate sequences of random bits, 0s and 1s. In the PEAR Lab tests, participants tried to intentionally influence the RNG outputs to drift *above* the chance-expected average (they called this the high aim condition, like aiming for heads instead of tails), then *below* chance (the low aim condition, like aiming for tails), and then to withdraw their mental intention entirely to allow the RNGs to behave normally as the baseline or *control* condition.

From their experiments, Jahn's team reached several conclusions. They found that in all of their experiments using truly random sources, like those based on quantum events, the random outputs tended to match the directions that the participants intended. When wishing for high scores the RNG outputs drifted up, and when wishing for low scores the RNG outputs drifted down. By comparison, no positive results were observed when simulated random numbers were used, like those generated by software algorithms. They estimated that the magnitude of the PK effect was approximately equal to 1 bit out of 10,000 being shifted away from chance expectation. While this may seem like a tiny effect, over the entire database this resulted in odds against chance of 35 trillion to 1 (Figure 9-3).[16]

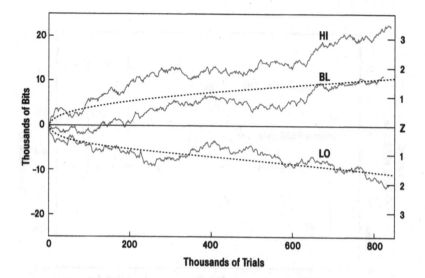

Figure 9–3. Results from 12 years of random-number-generator experiments at the PEAR lab. The HI, LO, and BL curves correspond to mentally aiming for high numbers (HI), low numbers (LO) and a baseline control (BL). The parabolas show the thresholds where effects exceed odds against chance of 20 to 1. The specific shape of the experimental curves is not as important as the fact that the three curves separate as predicted, according to the participants' mental intentions.

Three years after the review of the PEAR Lab RNG studies, a large scale or "mega-trial" experiment was jointly conducted by the PEAR Lab, the Institut für Grenzgebiete der Psychologie und Psychohygiene in Freiburg, Germany, and the Justus-Liebig-Universität Giessen, in Giessen, Germany.[17] This three-site consortium attempted to replicate the PEAR results using a similar design, similar equipment, and a preplanned series of trials. Though the replication attempt failed to provide a significant outcome, the results were significantly similar to the original PEAR findings (Figure 9–4), with odds against chance of 20 to 1.[18] Thus, while the outcome of the mega-trial was not independently successful in demonstrating PK effects, there was evidence that the same basic *trend* was repeated.

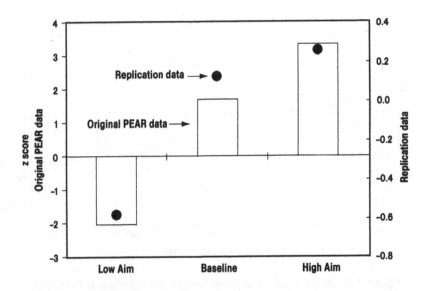

Figure 9–4. Deviations in random-number-generator outputs compared to chance expectation in the original PEAR data (white bars) and in data from a three-laboratory replication attempt (black circles). The magnitude of the original and replication results were different, but the trends in both cases were in alignment with the direction of mental intention.

Given the strong results in the original PEAR studies, and the similar outcomes in the mega-trial test, the question naturally arises as to whether this effect has been independently replicated by other investigators. In 1989, Princeton University psychologist Roger Nelson and I conducted a meta-analysis of all known RNG studies published up to that time. Updating that analysis again for this book, I found 490 studies comprising a total of 1.1 billion random bits subjected to PK intention.[19]

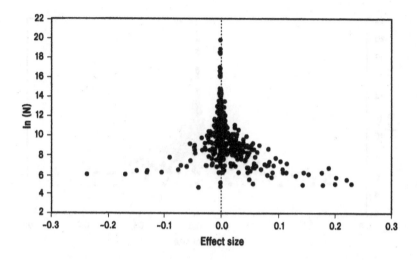

Figure 9–5. Funnel plot for 490 published RNG studies. The number of random bits used in these studies ranged from a few hundred to tens of millions, so to compress this range on the plot the *y*-axis is a log scale. The missing dots from the lower left portion of the plot suggests that this literature has a selective reporting problem.

The overall effect was small in magnitude, but associated with odds against chance of 50,000 to 1.[20] Selective reporting was a problem, as a bite was missing from the lower left side of the funnel plot (Figure 9–5), so I applied the trim and fill algorithm and estimated the number of potentially missing studies at 105 (Figure 9–6). In this worst-case adjustment for selective reporting, the overall level of significance remained significant with odds against chance of 3,050 to 1.[21]

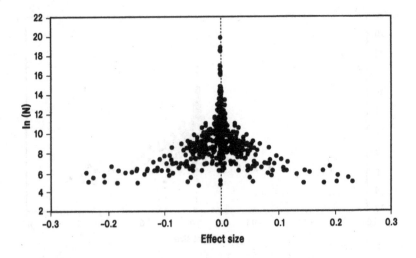

Figure 9–6. Revised funnel plot after adding 105 studies identified by the trim and fill method. The overall level of significance remained significant with odds against chance of 3,050 to 1.

Then I calculated the number of file-drawer studies required to nullify the existing result—it came to 2,610. This means that each of the 90 authors reporting at least one RNG study would have had to conduct an additional 29 nonsignificant studies, and failed to report any of them.[22] In a quality analysis, higher quality studies did not result in significantly lowered effects. So chance, selective reporting, and variations in study quality are—once again—not viable explanations for these results.

IS IT INFLUENCE?

If minds really can influence the outputs of RNGs, then our ability to detect that PK effect ought to improve as more bits are generated. To test this idea, I applied the same sort of analysis used earlier on the dice studies to the PEAR Lab RNG data. The resulting curve is remarkably similar to that observed with dice (Figure 9–7). This suggests that PK on random numbers also acts as a type of influence.[23]

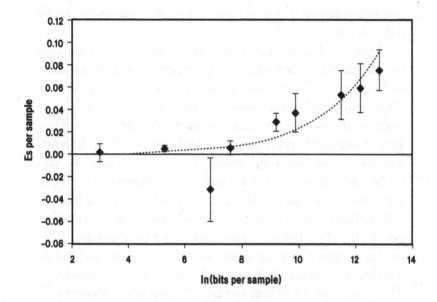

Figure 9–7. The dashed line shows the predicted increase in effect size with increasing (natural log of the) number of bits-per-button press. The diamonds with error bars show the actual results. This suggests that PK influences each bit about the same, analogous to a "mental force."

At this point, people often start conjuring up wondrous images of future-oriented psychic technologies. Venture capitalists clap their hands with glee as they imagine planes that fly where you will them to fly, cars steered by a flick of the mind, mentally controlled prosthetic devices . . . the possibilities are endless.

Ah, if only it were so simple. Unfortunately this PK-as-influence finding hasn't been solidly confirmed yet,[24] and some experiments with very high bit-generation rates have obtained highly significant results *opposite* to the intended direction.[25] This suggests there might be limitations on the speed with which random bits can be mentally influenced, or that there might be important differences between random events that are generated in parallel (dice) versus serially (RNG), or that PK might manifest differently in big physical objects like dice versus microscopic virtual objects like random bits. Much more basic research is required to turn these fragile phenomena into useful technologies.

These studies seem to imply that mind literally influences matter. But there are alternate interpretations. Perhaps mind and matter are like two sides of the same coin. To study such an effect, you could take a ribbon and write *mind* on the inside and *matter* on the outside. Now, as you wiggle the ribbon, you'll find very strong correlations between mind and matter, yet in a fundamental sense never the twain shall meet.[26] Then one day, while you're distracted for a moment, a mischievous friend cuts your ribbon, creates a half-twist, and carefully tapes it back together. Later you pick up the altered ribbon and proceed to ponder the abyss between mind and matter by absent-mindedly tracing a finger along the matter side of the ribbon. To your astonishment, you find that your finger ends up on the mind side! This is because the ribbon was transformed into a Möbius strip by your friend's half-twist, and this topological curiosity has only one side. The lesson is that sometimes simple twists in conventional concepts can unify things that appear to be quite different, like mind and matter. Some believe that consciousness may be the unifying "substance" from which mind and matter arirse, but defining one mystery in terms of another isn't particularly illuminating. At this point, all we can say is that when you begin to pry apart the mind-matter interface, it's as though that crack releases a dazzling and profoundly mystifying light. When you jostle the lever a bit more to boost the illumination, you encounter something even more mind-bending—effects that transcend time.

CHAPTER 10

PRESENTiMENT

Presentiment is that long shadow on the lawn
Indicative that suns go down;
The notice to the startled grass
That darkness is about to pass.

–Emily Dickinson

The term *presentiment* suggests a sense of foreboding, a vague feeling of danger, an intuitive hunch that something not quite right is about to unfold. Could such experiences involve perception of the future? One important hint that the answer is yes was provided in 1989 when Charles Honorton and Diane Ferrari published a meta-analysis of all "forced-choice" precognition experiments conducted between 1935 and 1987.[1] In a forced-choice test, a person is asked to guess which one of a fixed number of possible targets will be randomly selected later. The targets could be colored lamps, ESP card symbols, or the face of a tossed die. If the person's guess matches the randomly selected target, this is counted as a hit.

As in all psi experiments where the results depend on a well-defined value for chance expectation, the method of randomly selecting the future symbol is an important feature of these experiments. In the earliest studies, decks of cards were shuffled by hand or machine; in later studies, electronic RNGs were used to generate truly random numbers. The basic test is simple and the results are easy to interpret.

Honorton and Ferrari found 309 studies reported in 113 articles published from 1935 to 1987, contributed by 62 different investigators. The database consisted of nearly 2 million individual trials by over 50,000 subjects. The study designs ranged from the use of ESP cards to computer-generated, randomly presented symbols. The time interval between the guesses and the generation of the future targets ranged from milliseconds to a year. The combined results of the 309 studies produced odds against chance of 10^{25} to one. That's ten million billion billion to one, eliminating chance as an explanation. The possibility of a file-drawer problem was rendered implausible by determining that the number of unpublished, unsuccessful studies required to eliminate the observed results was 14,268.[2] Further analysis showed that 23 of the 62 investigators (37%) had reported successful studies, so the overall results were not due to one or two wildly successful experiments. In other words, the precognition effect had been successfully replicated across many different experimenters.

A decade later, philosopher Fiona Steinkamp and psychologists Julie Milton and Robert Morris, all from the University of Edinburgh, published a meta-analysis of forced-choice experiments comparing clairvoyance (perception of hidden targets in the present) with precognition (perception of targets in the future).[3] In 22 studies published from 1935–1997, they found overall significant evidence for both clairvoyance, with odds against chance of 400 to 1, and for precognition with odds of 1.1 million to 1. There was no difference in the magnitude of effects between these two modes of perception,[4] and there wasn't any evidence that these effects were explainable as methodological problems or procedural errors. Their conclusion was that psi works just as well for perception of real-time events as it does for future events.

UNCONSCIOUS PRECOGNITION TESTS

While forced-choice tests continue to generate interesting results, like most guessing tests they tend to produce very small effects that decline over time, probably due to the incredibly boring nature of the forced-choice tasks. To overcome these limitations, investigators began to explore unconscious forms of precognition. One of the earliest (1946) to suggest this was A. J. Good, brother of the well-known British statistician I. J. (Irving John) Good, who wrote about it in the *Journal of Parapsychology* in 1961:

> A man is placed in a dark room, in which a light is flashed at random moments of time . . . . The man's EEG (electroencephalogram) is recorded on one track of a magnetic tape, and the flashes of light on another. The tape is then analyzed statistically to see if the EEG shows any tendency to forecast the flashes of light.[5]

While Good's specific experiment hasn't been conducted yet, a number of studies resemble it. In 1975, Jerry Levin and James Kennedy used a reaction time task to see whether a slow brain-wave indicator of anticipation, called contingent negative variation (CNV), would unconsciously detect a stimulus that appeared in the future at a random time.[6] Participants were asked to anticipate the appearance of a green light, and then to press a key if a green light appeared but not if a red light appeared. An electronic RNG determined which color light would appear. As predicted, anticipatory brainwaves were observed just before the RNG selected a green light, as compared to a red light. A few years later, John Hartwell reported a similar study, using the same measure of anticipation.[7] He found that 13 of 19 planned tests were in the predicted direction, but overall he didn't find a significant result. An attempted replication by Hartwell, reported the following year, was also unsuccessful.[8]

Around the same time, Hungarian physicist Zoltan Vassy reported an experiment based on skin-conductance responses in an unusual type of telepathy experiment. In Vassy's study, two people were isolated in separate rooms. At random times, the sender received an electrical shock; 3.5 seconds later, the receiver also received a shock. The skin conductance of the receiver during the 3.5-second interval immediately preceding his shock was examined to see if it might rise due to telepathic anticipation of the up-coming shock. Five sender-receiver pairs took part in ten experimental sessions, of which six resulted in significant results, each with odds against chance greater than 100 to 1.[9] This was an amazingly strong result.

So amazing that these observations were soon forgotten. Then in the winter of 1993, while I was working at the University of Edinburgh and thinking about ways to improve the reliability of psi experiments, I envisioned a simpler way to test for the presence of unconscious precognition. I would monitor a person's skin conductance before, during, and after viewing emotional and calm pictures, and then see if the autonomic nervous system responded appropriately *before* the picture appeared. A few years later, I started running a series of experiments based on this design.

PRESENTIMENT

In this experiment, a participant (Jack) is asked to sit in front of a blank computer screen. I attach electrodes to his palm to record tiny fluctuations in skin conductance and then ask Jack to hold a computer mouse in his other hand. When he's ready to begin a trial, he presses the mouse button and waits for a picture to appear on the computer screen (Figure 10–1). After the button press, the computer waits 5 seconds, selects a picture at random from a large pool of images, displays it on the screen for 3 seconds, then it disappears and the screen goes blank again for 10 seconds. After that a message appears informing Jack to start the next trial whenever he's ready by pressing the mouse button again. This sequence is one trial in the experiment. Skin conductance is continuously monitored while Jack repeats 30 to 40 such trials in one session. The images he sees are either *calm* photos, such as landscapes, nature scenes, or calm people, or *emotional* photos, such as erotic, violent, or accident scenes.

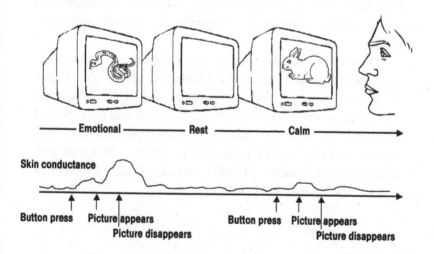

Figure 10–1. While skin conductance is monitored, the participant presses a button. Five seconds later, the computer makes a random decision to display either an emotional or a calm picture. Presentiment manifests as a rise in skin conductance before emotional pictures, but not before calm pictures.

The idea of presentiment assumes that we are constantly and unconsciously scanning our future, and preparing to respond to it. If this is true, then whenever our future involves an emotional response, we'd predict that our nervous system would become aroused before the emotional picture appears. If our future is calm, we'd expect to remain calm before the picture appears. Of course, *after* an emotional or calm picture appears the response is well understood as the "orienting reflex." This is the body's predictable reaction to a novel stimulus, in which it momentarily tenses up while evaluating whether to fight or flee.

A more general prediction of presentiment is that the body responds in advance of a future event *in proportion* to how emotional that future event will be. Extremely emotional future events will produce larger responses (before the picture appears) than mildly emotional future events. Likewise, extremely calm events will produce smaller responses than moderately calm events.

Twenty-four people participated in the first set of presentiment experiments I ran at the University of Nevada (Figure 10-2).[10] As expected, skin conductance reacted 2 to 3 seconds *after* the presentation of an emotional stimulus, and the expected difference between the calm and emotional responses was clearly evident. But the presentiment effect, which was predicted to occur *before* the stimulus, was also observed with odds against chance of 500 to 1.[11]

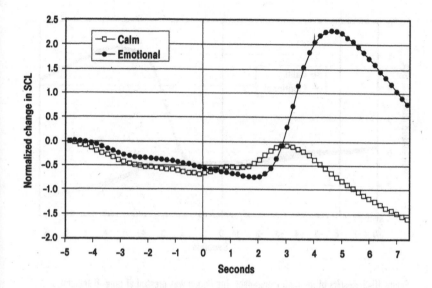

Figure 10–2. Results of my first set of presentiment experiments, showing changes in skin-conductance level before and after randomly selected calm and emotional trials. The vertical line in the graph, at time 0, shows when the randomly selected picture was displayed. The presentiment effect is the difference in the curves before time 0. In this case it was associated with odds against chance of 500 to 1.

In the second experiment, I ran 50 volunteers at the University of Nevada and 6 more at Interval Research Corporation, in Palo Alto, California. The results were in the predicted direction, but weren't as strong as those observed in the first experiment. The third experiment used new hardware and software, a new picture pool,[12] and a new group of 47 participants. In this study, the trial-initiating button press occurred 6 seconds before the stimulus (Figure 10–3) rather than 5 seconds in the prior experiments. The skin-conductance levels were virtually identical before the button press, but as soon as the button was pressed they began to diverge in accordance with the future stimulus. This study resulted in a strong presentiment effect, with odds against chance of 2,500 to 1.

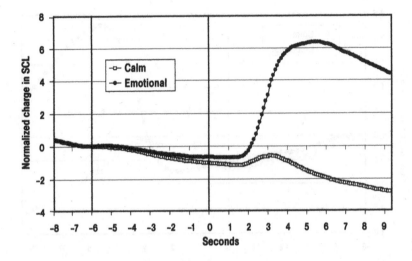

Figure 10–3. Results of my third experiment. The button was pressed at time -6 seconds, and the picture was randomly selected and displayed at time 0. The presentiment effect in this study was associated with odds against chance of 2,500 to 1.

Participants in the fourth study were recruited to test a new type of skin-conductance monitor. The results were in the predicted direction, but weren't statistically significant. Overall, however, the combined odds against chance for these four experiments was 125,000 to 1 in favor of a genuine presentiment effect.[13] These studies suggest that when the average person *is about to see* an emotional picture, he or she will respond before that picture appears (under double-blind conditions).

Recall that presentiment predicts that the prestimulus responses will increase as the emotionality of the future photos increase, indicating that specific information about the emotional content of the future image is perceived in the present. The correlation observed in these experiments was, as predicted, significantly positive, with odds against chance of 125 to 1.[14]

Presentiment is maddening to some scientists because it challenges commonsense beliefs about causality and time. Many philosophers also dismiss precognition as an incoherent concept suitable only for burning because it raises the specter of a logical paradox. So I knew that to provide a persuasive case for this evidence every conceivable loophole would have to be carefully examined and tightly closed. Alternative explanations could include sensory or statistical cues about the upcoming targets, data collection errors, measurement or analytical artifacts, selective reporting biases, participant or experimenter fraud, or a variety of conscious or unconscious anticipatory strategies. In fact, we considered all of these factors in the process of designing, running, and analyzing these experiments, and none could explain the results. Still, the proof of the pudding in science is what happens when other investigators try to repeat an experiment. The bottom-line question always comes down to: Is this a repeatable effect?[15]

REPLICATIONS

From 1998 through 2000, I directed a psi research program at Interval Research Corporation in Silicon Valley. Interval was Paul Allen's (cofounder of Microsoft) consumer electronics research laboratory that ran from 1992 to 2000. David Liddle, a pioneer in the development of graphical user-computer interfaces, headed it for most of that time. With nearly 200 scientists and technicians on staff, Interval attracted numerous legendary figures in the technology world, including Rob Shaw, cocreator of chaos theory; Max Mathews, the first person to make a computer play a tune; Joy Mountford, head of Apple Computer's user interface group; Jim Boyden, inventor of the inkjet printer; Richard Shoup, Academy Award recipient for co-developing the computer graphics techniques now used in movie special effects; and many other inventors from Xerox PARC, Apple, Stanford, Bell Labs, IBM, and the MIT Media Lab.

One of the research projects I conducted at Interval was on presentiment. At one point, I had an opportunity to demonstrate the experiment to Nobel laureate Kary Mullis, who was visiting Interval. A few weeks later, he appeared as a guest on National Public Radio's *Science Friday* program (May 1999). As part of that interview, Mullis described his experience as a participant in this experiment. He said, on the air, that we had demonstrated the presentiment effect to him: "I could see about 3 seconds into the future," he stated. Then he added,

> It's spooky. You sit there and watch this little trace, and about three seconds, on average, before the picture comes on, you have a little response in your skin conductivity which is in the same direction that a large response occurs after you see the picture. Some pictures make you have a rise in conductivity, some make you have a fall. He's done that over and over again with people. That, with me, is on the edge of physics itself, with time. There's something funny about time that we don't understand because you shouldn't be able to do that . . ."

PRESENTIMENT BEYOND THE HUMAN

Soon after this NPR program aired, I was contacted by Chester Wildey, a master's-degree candidate in electrical engineering at the University of Texas at Arlington. He heard the radio interview with Dr. Mullis, was intrigued, and convinced his thesis committee that, although the presentiment hypothesis was unorthodox, it had attracted the interest of a Nobel laureate. So he got permission to design and build a skin-conductance monitoring circuit and test it in a presentiment experiment.

Wildey tested 15 participants with a total of 314 trials. He noted in his thesis that he thought precognitive phenomena might be feasible if the "quantum mind" possibilities proposed by University of Arizona anesthesiologist Stuart Hameroff and Cambridge University mathematician Sir Roger Penrose had any merit. As he put it,

Dr. Hameroff and Penrose's theory of mind predicts that consciousness should occur in brains down to the size found in worms. Since change in skin impedance is thought to be related to internal mental state, an interesting question is whether responses such as those in the above experiment could be observed in lesser species. With this in mind, an additional series of tests was conducted on earthworms.

Wildey tested earthworms in 231 trials. In 114 of them he used a mechanical vibration for the earthworm equivalent of an emotional stimulus, and in 117 he used no vibration as a control. Wildey found that the results of both tests were in alignment with results reported in my experiments, and the combined human and earthworm results were very nearly statistically significant (odds against chance of 17 to 1).[16]

Wildey also found that the more trials he collected, the closer his data agreed with the presentiment hypothesis, which is what one would expect if the "signal" is a genuine one. Then, to confirm that his equipment had not accidentally introduced a mistake that looked like the presentiment effect, he designed a circuit that simulated fluctuations in human skin conductance, ran the "sham human" circuit through the experiment, and obtained results that stood right at chance. Wildey concluded that

> The results of this experiment support the hypothesis that skin impedance changes predict randomized future emotional responses in people . . . . Similar results were found using earthworms with the time window extending one second before the stimulating event compared to three seconds before the event in human subjects.

PRESENTIMENT IN THE HEART

In 2004 in the *Journal of Alternative and Complementary Medicine*, psychophysiologist Rollin McCraty and his colleagues reported a presentiment experiment using skin-conductance, heart-rate, and EEG measures.[17] The study had two experimental conditions: before meditating and after meditating for 15 minutes. The meditative state was based on a self-regulation training scheme called "Freeze Frame," which includes a breathing and visualization exercise said to induce a state of mind/body resonance within the body.

McCraty's experiment closely followed the basic presentiment design. He ran 26 adult participants and obtained positive (but not significantly so) evidence for presentiment in skin conductance. His results with heart and brain signals were more interesting. He found that heart rate significantly slowed down before future emotional pictures as compared to calm pictures, with odds against chance of 1,000 to 1 (Figure 10–4), that women performed better than men, and that the brain responded differently before emotional and calm stimuli. He summarized his findings as follows:

> Of greatest significance here is our major finding: namely . . . evidence that the heart is directly involved in the processing of information about a future emotional stimulus seconds before the body actually experiences the stimulus . . . . What is truly surprising about this result is the fact that the heart appears to play a direct role in the perception of future events; at the very least it implies that the brain does not act alone in this regard.[18]

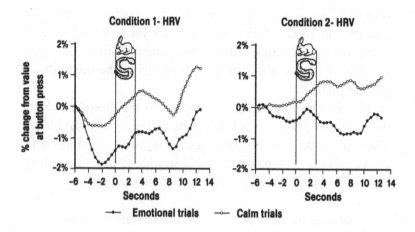

Figure 10–4. Presentiment effect associated with variations in heart rate (HRV means "heart-rate variability"), as reported by Rollin McCraty and his team. The calm condition (rabbit picture) refers to the top curve, the emotional condition (snake) refers to the bottom curve. Condition 1 was prior to meditation and Condition 2 after the meditation. Both conditions produced highly significant differences in heart rate as predicted by the presentiment hypothesis.

PRESENTIMENT AND PERSONALITY

In another presentiment study conducted in 2004, psychologist Richard Broughton from the University of Northampton, England, gave participants in a presentiment experiment two questionnaires to see what role personality might play in the outcome. He also tested each participant twice to see if their performance in the two sessions would be similar.

One of the questionnaires was the popular Myers-Briggs Type Indicator (MBTI). This personality test was used because its intuition and extraversion scales have been found to correlate with psi performance in previous laboratory tests.[19] The second questionnaire was the NEO-Five Factor Inventory (NEO-FFI). While not as well known as the MBTI, the NEO-FFI is a standard survey used in personality research, and its "openness" factor has been found to correlate with better psi performance. Broughton predicted that these personality factors might also show a positive relationship with presentiment.

He ran a total of 128 people in his test and found results in the predicted direction, but they weren't statistically significant. Then he looked at the relationship between the individual scores and the personality tests. He concluded that of the three personality scores predicted to show better performance,

> All three were correlated in the expected direction and two of them, MBTI Intuition and NEO-FFI Openness, were correlated significantly with [presentiment] . . . . This result is a promising indication that experiments with more robust evidence of [presentiment] may reveal personality relationships consistent with previous ESP research.[20]

FREE-RUNNING PRESENTIMENT

In 2003, physicists James Spottiswoode and Ed May reported a presentiment experiment with two new twists. Their experiment used audio instead of images and employed a "free-running" design, instead of asking participants to initiate each trial at will. When a person ran this experiment, they simply relaxed for about 30 minutes while randomly, about once a minute, they'd hear a very loud, one-second sound blast over headphones, or they'd "hear" one second of silence as a control. This design is closer to real-life intuitive hunches in that the participant doesn't initiate anything—each trial just happens automatically, at random times.

Spottiswoode and May predicted that their participants would show more fluctuations in skin conductance before the audio stimulus than before the control moment of silence.[21]

After running 125 volunteers through the experiment, their prediction was confirmed with odds against chance of 1,250 to 1 (Figure 10–5).

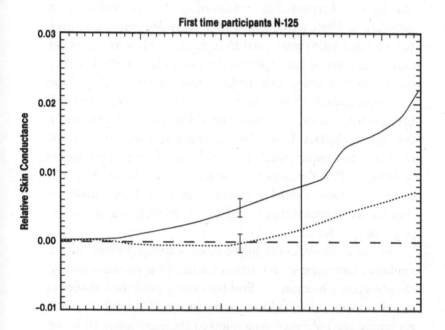

Figure 10–5. James Spottiswoode and Ed May's audio-presentiment experiment, with 125 participants. The top line refers to the audio condition, the bottom to the silent condition. The top curve shows skin conductance (and one standard error bar) just before a randomly timed one-second audio tone, and the bottom curve shows skin conductance before a randomly timed one-second moment of silence, as a control. The difference between the curves is associated with odds against chance of 1,250 to 1, in alignment with the presentiment hypothesis.

As in all of the other presentiment studies, they examined whether anticipatory strategies might explain these results and found that they could not. Because of the success of this approach, May asked a colleague, physicist Zoltan Vassy of Budapest, Hungary, to try replicating this free-running design.

Vassy ran 50 new participants, and again obtained a significant result, with odds of 20 to 1.[22]

BIERMAN'S BRAIN

Shortly after I reported the results of the first presentiment experiment in 1996, psychologist Dick Bierman from the University of Amsterdam attempted to replicate it. He was successful, and since then he has repeated the presentiment effect numerous times.[23] Bierman soon realized that this effect, if genuine, must also appear in mainstream psychophysiological research because the experimental method used to study presentiment is not unique. In fact, I modeled the original design after an elementary technique used by academic psychophysiologists worldwide. But if presentiment is really so common, then why hadn't someone noticed it before? The most likely answer is that no one expected that it might exist, so there was no reason to go looking for it.

To see if presentiment might have been overlooked in conventional experiments, Bierman examined the mainstream psychophysiology literature to find previously published studies of emotion using skin-conductance measures. He found two studies where the published data allowed the experiment to be reconstructed into a test for presentiment. The first was a study of gambling behavior in brain-damaged and normal people. The second was concerned with the speed with which fear arises in people afraid of animals like spiders or snakes.[24] Bierman asked an assistant to extract the data from those papers without telling her what he was interested in. To his amazement, the combined data produced significant presentiment effects, with odds against chance greater than 100 to 1.[25]

This outcome, which suggests that presentiment effects might be more common than previously supposed, motivated Bierman to conduct an especially intriguing presentiment experiment. He used a functional magnetic resonance imaging (fMRI) system, which measures the amount of oxygen in the blood, to see *where* in the brain the presentiment effect would appear.[26] The acronym for a common fMRI measurement is BOLD, which means Blood Oxygenation Level Dependent. The fMRI is noninvasive, meaning nothing is injected into the person, and it allows relatively fast-changing events in the brain to be observed as they happen.

The idea behind the use of the BOLD measure is that areas of the brain that are more active have higher levels of oxygenated blood as compared to less active portions of the brain. While the fMRI can measure the BOLD values in 100 millisecond time slices, blood can't move through the brain that quickly. It generally takes a few seconds for an fMRI to measure noticeable differences, and thus in an fMRI experiment a person is asked to perform one mental task for a few seconds, and then switch to another task, and then repeat that cycle. The goal is to find the areas of the brain that use more oxygen while engaged in the first task as compared to the second.

Bierman and his colleagues designed an experiment in which participants located inside an fMRI were asked to look at computer-projected images. After each picture they were asked to remain as calm as possible, to not think about the pictures they had already seen, and to avoid anticipating the upcoming pictures. The pictures in his test included 18 erotic, 18 violent, and 48 calm images. The pictures were selected at random on each successive trial. Each trial began with the participant looking at a fixed point on an otherwise blank screen for 4.2 seconds, then a picture appeared for 4.2 seconds, and then the picture disappeared and the trial continued with a blank screen for 8.4 seconds.

To test the procedure, Bierman first ran the experiment with himself as the subject (Figure 10–6). He found a presentiment effect in the BOLD measures before the erotic pictures, with odds against chance of 320 to 1.[27] But he showed no presentiment effect with either violent or neutral images.

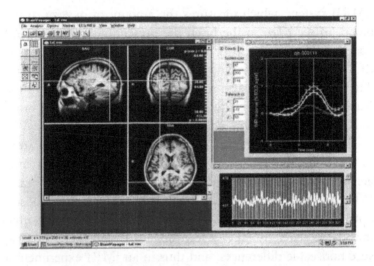

Figure 10–6. Upper left: Crosshairs show where a presentiment effect appeared for erotic images in Dick Bierman's brain. Upper right: Average blood oxygenation levels (BOLD) values from –8 to +12 seconds; the stimulus presentation occurred at t=0. Lower right: BOLD signal over the course of the entire experiment in the region indicated by crosshairs in the upper left images. The rise in BOLD before the erotic images was associated with odds against chance of 320 to 1.

Bierman was encouraged by this result, and noted: "It should be stressed that this is a typical result and not due to a long search for a region that would show this result."[28] Given the positive outcome, he decided to run 10 adult volunteers in a more formal test. When all the data were in, he examined the data from males and females separately because he expected that the responses to the emotional pictures might depend on gender. Then he ran an analysis identical to the one he used on his own brain.

The results showed presentiment effects in most of the individual brains, and it was widely distributed over many brain regions. After averaging across all the participants, significant differences appeared in one common brain area. For females there was a significant presentiment effect for erotic images (odds against chance of 25 to 1) and for violent images (odds of 50 to 1). For males there was no difference for the violent images, but there was for erotic images (odds of 50 to 1).

Lest we forget what's going on in this experiment, it's useful to be reminded what these results mean: The brains of both men and women were activated in specific areas *before* erotic pictures appeared, even though no one knew in advance that those pictures were about to be selected. In other words, *the brain is responding to future events.*

Given the controversial nature of this claim, Bierman discussed in detail alternative explanations for these results. This included the possibility that the results were due to "fishing" for good results in the data, that the results were due to chance, and many other conceivable explanations. He concluded that the fMRI results were valid, and in agreement with the other studies based on skin-conductance and heart and brain measures.

LOOKING INTO THE FUTURE, SO FAR

Presentiment experiments provide a new form of evidence suggesting that we can unconsciously perceive our future. How far into the future we might be able to sense remains uncertain (as does the meaning of "the future"). Like most psi effects, the results in these studies are relatively small in magnitude, but they appear in a wide range of people tested and they are consistent across many different types of tasks, measurements, and personality types. These effects even appear in experiments conducted for other purposes.

When you step back from the details of these studies, what you find is a spectacular body of converging evidence indicating that our understanding of time is seriously incomplete. These studies mean that some aspect of our minds can perceive the future. Not infer the future, or anticipate the future, or figure out the future. But actually *perceive* it.

While you ponder this, hold on to your socks because we're not finished.

So far, the repeatable laboratory evidence suggests that we have the capacity to perceive distant information and to influence distant events, across space and time. This data challenges the assumption that we are isolated creatures, separated in space and time, and it implies that our intentions may not be limited only to own minds and bodies. If all this is true, and if intention and attention are "spread out" in space-time, then the question arises, do individual intentions at times coalesce into group intentions? And if so, what effects might such "group minds" have?

CHAPTER 11
GAIA'S DREAMS

Peoples and civilizations reached such a degree either of frontier contact or economic interdependence or psychic communion that they could no longer develop save by interpenetration of one another. . . . Under the combined influence of machinery and the super-heating of thought, we are witnessing *a formidable upsurge of unused powers.*[1]

—Pierre Teilhard de Chardin

When Jesuit priest, paleontologist, biologist, and philosopher Pierre Teilhard de Chardin wrote these words in 1955, ideas like global warming, multinational corporations, and worldwide digital networks were fantasies. But Teilhard de Chardin saw farther into the future. He envisioned the "noosphere," a planetary consciousness. Was this vision of Gaia, the Greek name for the Earth goddess, a premonition, or was it wishful thinking?[2]

In entertaining this question, we are obliged to reconsider the limits of psi. Are we dealing with a personal perceptual skill that can be used to our private advantage, like an improved form of visual acuity or more sensitive hearing? Do our minds occasionally dip into a holistic reality and bring back tidbits of useful information? Or is something bigger going on, something that transcends the individual altogether? One only needs a minor leap of imagination to imagine that if psi is real–and given the experimental results, that seems an increasingly safe bet–then in the same way that networks of neurons combine to form our brains, maybe psi forms an interconnective web of brain/minds that results in a collective mind. But if so, how would we test this idea?

One way is by using truly random number generators (RNGs). We do this because we know from previous laboratory tests that mind-matter interactions can be detected in the behavior of RNGs. These devices are also convenient because they can be programmed to automatically run quietly in the background as passive "observers" of the proposed collective mind-matter interactions. Princeton psychologist Roger Nelson initiated this type of research, called "field consciousness" experiments, in the mid-1990s.

These studies rely on the fact that RNGs are designed to generate pure randomness, technically known as *entropy,* and that fluctuations in entropy can be detected using simple statistical procedures. If it turns out that the recorded entropy decreases when one of these random generators is placed near groups engaged in high focused attention, like a group meditation or a deeply engaging spiritual ritual, then we can infer something about the presence of coherent minds possibly infusing the environment with an ordering "field" that reduces entropy. In other words, if we assume that mind and matter are related, then when one side of the mind ↔ matter relationship changes by becoming highly ordered, the other side of the equation should show unusual forms of order as well.

FIELD-CONSCIOUSNESS EXPERIMENTS

By 2005, more than a hundred field-consciousness experiments had been reported by groups in the United States, Europe, and Japan. This included experiments conducted at Native American rituals, popular festivals in Japan, theatrical performances, scientific conferences, psychotherapy sessions, sports competitions, and live television broadcasts.[3] Overall, these studies strongly suggest that coherent group activity is associated with unusual moments of order in RNG outputs.

Engineer William Rowe was intrigued by these studies because during creative brainstorming meetings he sometimes felt palpable moments that were permeated with "focused group energy." As he put it:

> Every so often we hear of a group of people who unite under extreme pressure to achieve seemingly miraculous results. In these moments human beings transcend their personal limitations and realize a collective synergy with results that far surpass expectations based on past performance. Anyone hearing a fine symphonic or jazz group hopes for one of those "special" concerts that uplift both the audience and the performers. Perhaps less frequent, but often more spectacular, are examples in sports, such as the 1980 U.S. Olympic Hockey Team, a group of talented amateurs who stunned the world by winning the gold medal against the vastly more talented and experienced, virtually professional Russian and Finnish teams. These occurrences, although unusual, are much more frequent in American business than is commonly suspected.[4]

Rowe wondered whether a group experiencing one of those synergistic moments would influence a nearby hidden RNG. So he designed an experiment to test this idea. On the mind side of the mind-matter equation, a positive *subjective* result meant that during a brainstorming session an observer sensed that periods of focused group energy were occurring. A positive *objective* result meant that the RNG outputs significantly deviated from chance for at least one minute during the meeting. If both of those events occurred at the same time in a given meeting, it was counted as a "true positive." Likewise, a "true negative" meant that the observers did *not* report a sense of group coherence and the RNG results were in alignment with chance. A "false positive" meant the observers reported group coherence but the RNG did not, and vice versa for "false negative."

Rowe conducted 11 formal tests investigating the subjective and objective sides of group coherence. Each of these sessions was planned in advance, RNG data were taken before, during, and after each session, and an observer recorded his or her impressions about each session before the RNG outputs were examined. The 11 experiments resulted in the following:

	Observer report positive	Observer report negative
RNG positive	8	0
RNG negative	0	3

In other words, in all 11 tests the observer's impressions and the RNG outputs correctly matched. Rowe concluded that the field-consciousness experiment "seems to be a reliable detector of the coherent mind focus of groups of people as opposed to individuals working alone . . . . Empirical evidence in the form of single blind experimental protocols provides direct evidence that episodes of Focused Group Energy occur, and are both sensed by people and are physically measurable."[5]

Of course, not every field-consciousness study will produce a successful outcome. In reviewing the results of many of these studies, both those that worked and those that didn't, Roger Nelson was able to develop a recipe for contexts that seem to provide the most positive results. The recipe involved times and places that evoke unusually warm or close feelings of togetherness, with emotional content that tends to draw people together, where personal involvement is important but focused more toward a group goal involving a deeply engrossing theme, located at uplifting physical sites like the ocean or mountains, during creative or humorous moments, and enlivened with a sense of freshness or novelty.[6] By contrast, the opposite contexts—where individuals are working alone, are involved primarily in objective and analytical tasks, where there's low personal involvement and meaning, and where the tasks are boring or tedious—tend not to produce field-consciousness effects.

HEALING ENVIRONMENTS

One context that closely matches Nelson's recipe is the practice of *intentional healing,* or holding a focused desire for another person to achieve or sustain a state of health,[7] through such alternative healing methods as Qigong, Reiki, and Therapeutic Touch. Field-consciousness studies published in such settings have shown significant changes in RNG outputs.[8] But does this mean that coherent states of mind literally influence the environment?

Not necessarily. What these studies actually reveal is that intention and RNG outputs are *correlated,*[9] and correlation doesn't imply causation. Why? Because the movement of sunflowers over the course of the day is closely correlated to the apparent movement of the sun, but sunflowers do not *cause* the sun to move (or the Earth to turn). In this case, the mistake is in the assignment of the *direction* of causation, flower→sun. In this simple example we know the direction of causation, but in trickier circumstances like mind-matter interactions, figuring out what causes what isn't so obvious.

After running many such field-consciousness experiments, and observing these correlations appearing more often than not, I became increasingly interested in whether these coherence effects were actually *caused* by mind. One way to explore this question is to see whether mind-matter interactions correlated with changes in a nonliving system (like an RNG) would also simultaneously correspond to changes in a living system. If two different kinds of physical systems "responded" at the same time, a causal link might exist.[10] To develop this experiment, I worked with molecular biologists Ryan Taft and Garret Yount from the California Pacific Medical Center (CPMC) Research Institute in San Francisco.

For our living system we used cultures of astrocytes, the most abundant cell type in the human brain. We wanted to see whether cultures of those brain cells would grow more when exposed to healing intention as compared to when they were not exposed. We used these living cells as targets of healing intention because presumably individual cells don't "care" whether someone is trying to heal them or not, so they provide a rigorous way of testing the effects of healing intention while bypassing the problems of placebo effects (expectation) that complicate studies involving humans. For our nonliving targets system, we used three truly random RNGs, each based upon a different type of random source.

Our study also explored the idea that healing intention practiced repeatedly at the same location might change the physical site itself into a healing location. With sufficient exposure, such sites are thought by some to generate healing properties similar to those produced by a healer. This idea is weakly supported by stories of spontaneous healing at religiously numinous locations such as Lourdes and other sacred sites.[11] It's also supported by the concept of "place memories," or physical and psychological sensations repeatedly reported at certain locations, like traditionally haunted sites.[12] Of course, anecdotal reports about strange happenings at sacred or haunted locations are biased by expectations that something spooky might occur, so this is why we brought the phenomenon into the lab to study it under controlled conditions.

Four experienced Johrei practitioners took part in a three-day experiment. Johrei is a spiritual healing practice founded in Japan by Mokichi Okada (1882–1955). As in many spiritual healing traditions, Johrei maintains that there's a universal energy or spiritual force that can be cultivated and directed by intention. When focused on the human body, Johrei is said to raise its "spiritual vibrations," or to achieve spiritual purification; this in turn is said to improve health and to allow one's divine nature to unfold. Johrei practice assumes that for optimal healing repeated treatments are required to help overcome the body's inertia. Johrei also assumes that healing intention is not limited exclusively to the body, but affects the physical surroundings where healing treatments are provided.

In preparation for each day of the three-day experiment, Ryan Taft placed cultured human brain cells into 16 sealed flasks, each flask containing a nutrient solution to keep the cells alive. He selected two flasks at random as one set of controls and left two more flasks inside a cell culture incubator as a second set. Then he put the remaining 12 flasks inside a thermally insulated box and drove them to the Institute of Noetic Sciences laboratory, about 40 miles away. At the lab, he stored the flasks in one room and periodically brought groups of three flasks into the electromagnetically and acoustically shielded chamber where the healing trials would take place (as the treatment condition) or not take place (as a control condition). Taft never knew whether a given session was a treatment or control session to ensure that he didn't accidentally or intentionally handle the cell flasks differently.

During a treatment session, a Johrei practitioner directed his healing intention toward the box from about two feet away. He continued doing this for 25 minutes, without touching the box, then left the chamber and waited to be called for another session. This process was repeated four times a day for each of three days, randomly alternating healing sessions with control sessions. Between each pair of healing sessions, four Johrei practitioners met together inside the shielded room to practice a chanting meditation and to give each other Johrei healing treatments. These sessions, which lasted an hour and a quarter, were designed by the healers to help "condition the space" of the shielded room to further enhance the healing treatments.

After each day's experiments, Taft returned all of the flasks back to the cell incubator in the lab in San Francisco. Ten days later, the cells in all flasks were "fixed" to stop further cell growth, and then stained to make it easier to identify the cell cultures. Two lab analysts not otherwise involved in the experiment independently counted the number of cell colonies in each flask.

The three RNGs used were designed to operate reliably regardless of variations in external conditions, including fluctuations in temperature, electromagnetic fields, vibration, and electronic component aging.[13] This means that variations observed in the RNGs outputs would not be due to any mundane causes. Long-term calibration tests confirmed that these particular RNGs generated data according to chance expectation.[14] Two RNGs were hidden behind a curtain inside the shielded room. The Johrei practitioners knew that the RNGs were present, but there wasn't any feedback provided about their ongoing operation.[15] A third RNG was a computer-monitored Geiger counter.[16] That device monitored background ionizing radiation, including ambient alpha, beta, gamma, and X-ray particles, in 10-second samples. The Geiger counter was located about six feet outside the shielded room in a hidden location. All three RNGs were run before, during, and after the Johrei healing sessions, ending up with about one million one-second samples for the two electronic RNGs and 100,000 samples for the radiation-based RNG.[17]

As predicted by the "conditioned-space" hypothesis, the treated cells grew more as the experiment progressed (Figure 11–1). The odds against chance for the increased growth trend in the treated cells was 1,100 to 1.[18] By contrast, the control cells did not show a significant trend.[19] This suggests that repeated exposure to Johrei resulted in increased brain cell growth.

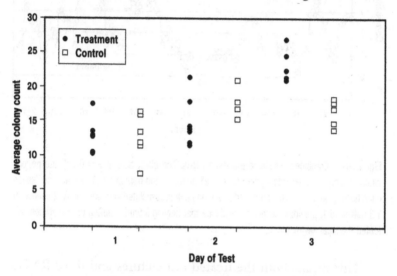

Figure 11–1. Average cell-culture counts for treated and control flasks, by day of test, excluding two treatment and two control colony-formation counts identified as unreliable. There were six treatment and control flasks per day, but some of the cell-colony counts were so close that the points on the graph overlap. This outcome suggests that the Johrei treatments caused the treated cells to grow more than the untreated control cells.

That was interesting, but the real surprise came when we examined the results of the three RNGs. We found that the three devices combined produced a peak response in the morning of the third day with odds against chance of 1.3 million to 1 (Figure 11–2).[20] Each of the three RNGs independently peaked at the same time.

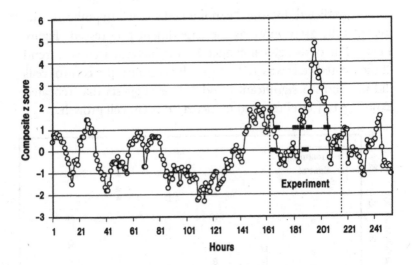

Figure 11–2. Combined results of the three RNGs. The black bars at z = 0 indicate periods when healing intention was applied to the cell cultures, and bars at z = 1 indicate the "space conditioning" activities. The combined deviation peak is associated with odds against chance of 1.3 million to 1, providing secondary evidence that healing intention during the experiment affected the local environment.

This means both the treated cell cultures and three RNGs significantly deviated from chance at around the same time on the third day. As usual, we considered a broad range of ordinary explanations for these results. This included chance, design flaws in the RNG hardware and software, natural environmental fluctuations, differences in how the treated and control cell culture flasks were handled, use of inappropriate statistics, selective reporting of data, and so on.[21] After much detailed investigation, we found that most of these alternative explanations were rendered implausible by the original experimental design, and the remaining explanations were eliminated by the observed outcomes. In short, this experiment suggested that certain forms of focused attention appear to causally *influence* both living and nonliving systems.

As soon as we observed the striking RNG results in this

experiment, we all thought, "What in the world was happening when those three RNGs in the lab 'spiked' on that third day? Did other RNGs, located elsewhere, also jump at the same time?" The answer would have told us something interesting about the role of distance in field consciousness effects. For a day or two we were sulking over our missed opportunity, and then I suddenly realized that there *were* random data available, and from the same kinds of RNGs. They were part of a worldwide network of RNGs called the Global Consciousness Project, which I'll describe later in this chapter.

I gained access to the data generated by 36 of those RNGs, located from 24 to 10,500 miles from our lab. These data allowed us to test three models: The first assumed that field consciousness effects are strictly *local* in the sense that RNGs that are near to the source of healing intention would show deviations from chance, but distant RNGs would not. The second assumed that these effects are *nonlocal* in the sense that intention is not bound by ordinary spatial constraints, and thus we'd expect to see large deviations in all of the RNGs at the same time, regardless of where they were located. And the third assumed that these effects would exhibit some sort of *distance-dependent* properties in the sense that deviations in RNGs would decrease with increasing distance from the laboratory.

To test these models, we compared measures of deviation in the three RNGs in the lab with similar measures calculated for 36 distant RNGs. The results showed strong evidence for the third model, distance dependence (Figure 11–3), with odds against chance of 37,000 to 1.[22]

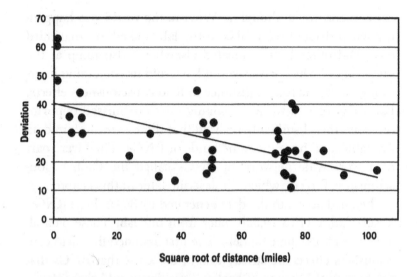

Figure 11–3. Deviations from chance in each RNG with increasing distance from the IONS laboratory. The effects of healing intention seems to decline with increasing distance, resembling a radiation-like effect.

Seeking to obtain further detail on this distance dependence, we decided to examine the combined deviations from chance in five of the RNGs located in the San Francisco Bay Area, all within 100 miles of the IONS laboratory, then to do the same for six RNGs located 6,000 miles away or farther. We were amazed to find that the group of five Northern California RNGs collectively peaked significantly above chance *at the same time* as the three RNGs in the laboratory. By comparison, the six distant RNGs hovered around chance. The combined effect for the five nearby RNGs and three RNGs in the lab was associated with odds against chance of 5 million to 1. This suggests that healing intention (at least as observed in this study) can act at a distance, but perhaps not at arbitrarily long distances.

One characteristic of psi, as revealed by both spontaneous reports and in lab tests, is that psi is not tightly bound to "now," in either space or time. But there is also evidence, like the results of this experiment, that psi may not be completely independent of distance. For example, in philosopher Fiona Steinkamp's analysis of the ESP card-guessing tests of J. B. Rhine and his colleagues, she examined the study outcomes according to distance between the person guessing the cards and the location of the cards themselves. She found a decline with increasing distance (Figure 11–4). Is this decline due to an inherent property of psi, or to the participant's knowledge that the target was at a distance? This remains an open question.

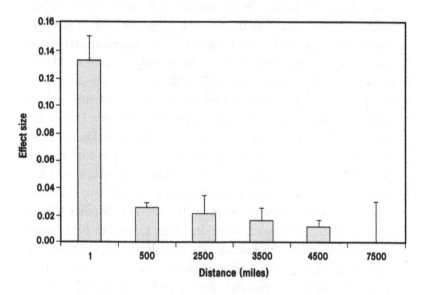

Figure 11–4. ESP card test results at different average distances, with one standard error bars. An effect size of zero in this graph is chance expectation. The decline in effect size with increasing distance suggests that psi effects may not be completely independent of distance.

PRINCESS DIANA

On August 31, 1997, an event took place that riveted the world's attention. Princess Diana and her companion Dodi al-Fayed were killed in a car accident in Paris. That tragic event saturated the world's news broadcasts for the next few days, and we all soon learned that Princess Diana's funeral would be broadcast live, worldwide, a week later. Those of us interested in the field-consciousness concept realized that Princess Diana's funeral would provide an interesting test case of *global* mental coherence, as hundreds of millions of viewers worldwide were expected to watch the funeral on live television.[23]

A dozen of us who had random generators, located throughout the United States and Europe, each ran our RNG before, during, and after the funeral. After the funeral we combined the outputs of the separate RNGs. Compounded across the twelve independent devices, we found a significant deviation, with odds against chance of 100 to 1, in alignment with our prediction of a global coherence effect. By unhappy coincidence, Mother Teresa died a few days after Princess Diana. Because we had just conducted the test for Princess Diana, most of us still had our RNGs in place, so we ran a similar experiment during Mother Teresa's funeral. That outcome wasn't significant. In thinking about the outcomes of the two tests, we realized that the contexts were quite different. Mother Teresa was 87 when she passed away and was known to be in poor health. Also, while her funeral was broadcast live, the proceedings were held in several languages without translation, and at times the television pictures were garbled or lost altogether. Those technical problems, combined with the different context of Mother Teresa's death, may have reduced the degree of highly focused attention that Princess Diana's funeral had attracted.

In any case, the success of the Princess Diana experiment and failure of the Mother Teresa experiment persuaded us that "global mind" experiments were worthwhile pursuing. For pragmatic reasons, the RNGs needed to be running continuously and automatically, and they had to be located in many places around the world. The idea would be to use this system to conduct field-consciousness experiments, to see, in other words, whether large scale coherence was generated during planned events, like New Year's Eve celebrations, but it could also be used for unexpected events, like the tragic deaths of celebrities, natural disasters, and terrorist attacks.

In late 1997, Roger Nelson took up the challenge, and with assistance from John Walker, the founder of AutoDesk, the computer-aided design company, and computer scientist Greg Nelson, devised a clever architecture to support an Internet-based, worldwide, continuously running field-consciousness experiment.

THE GLOBAL CONSCIOUSNESS PROJECT

The Global Consciousness Project (GCP), so dubbed and directed by Roger Nelson since its inception, significantly expands the one-shot field-consciousness experiment of earlier years. Instead of inferring coherence in a small group of people engaged in a common event, the GCP allows us to infer periods of global mental coherence as a result of major news events that attract widespread attention. With the advent of instant worldwide media and a growing number of internet-based news alerts services, the GCP postulates that within minutes of such major events, a sizeable percentage of the world's population will learn about those events, and as a result of the shift in global attention and the accompanying mental coherence, the RNGs located around the world would also begin to deviate from chance behavior.

How, you wonder?

Imagine a vast, windswept ocean with scores of buoys dancing in the waves. Each buoy has a bell attached to it to alert passing ships about hidden reefs and shallows. The sounds of each buoy's bell are broadcast by radio to a land-based central receiving station. This station receives the transmissions and consolidates them to form a single collective tone reflecting the ocean's grand dance. Most of the time this sound is unpatterned, similar to the random tinklings one might hear from a set of wind chimes dangling in a breeze. But every so often these buoys, isolated from one another by thousands of miles, mysteriously synchronize and swell into a great harmonic chord. When this occurs we know that something big has affected the entire ocean.

Because buoys reflect only wave surfaces, and the ocean is complex and deep, most of the time we can only guess at what caused the big event. One possibility is an underwater earthquake, like the event that spawned the tragic Asian tsunami of December 26, 2004. Another possibility is that a meteorite hit the ocean. A third explanation, one closer to the present topic, is that something subtle stirred in the ocean's depths, possibly quite delicate at its origin, but powerfully encompassing the entire ocean after rising from the deep.

Whatever the ultimate cause, we are interested in two types of analyses when the random bell tones spontaneously coalesce into a great chord. The first is how loud it is (the amplitude of the tone), the second is how coherent it is (the degree of harmony of the tone). The GCP is thus analogous to monitoring the ocean's surface for the presence of a tsunami. Except instead of looking for massive movements of water and inferring what is happening in the depths of a great ocean, we monitor massive movements of entropy generated by a network of RNGs and infer what is happening in the depths of a "grand mind."

Each RNG in the GCP network is attached to a computer that collects one sample (of 200 bits) per second. (The sources of randomness in the RNGs include electronic noise in resistors and quantum tunneling effects in diodes.) Each computer records its trials into time-stamped files, and all computer clocks are synchronized to standard Internet time. Every five minutes, all data are automatically assembled and sent over the Internet to a central web server in Princeton, New Jersey.

The Global Consciousness Project's network began with three RNG sites in 1998. Over time, it increased as volunteers were found who were willing to host an RNG on their personal computer. By April 2005, the network included about 65 active RNGs located mostly throughout Europe and North and South America, but also in India, Fiji, New Zealand, Japan, China, Russia, Africa, Thailand, Australia, Estonia, and Malaysia.

The global mind-matter interaction hypothesis is tested in the data by examining whether the streams of random bits generated by the RNG network change from chance expectation in predefined ways. For most events, this analysis examines RNG data from a few minutes before an event of interest occurs to a few hours afterwards. By April 2005, a total of 185 events of global interest had been tested and double-checked by independent analysts. These events included new years celebrations, natural disasters, terrorist activity, massive meditations, sports events, outbreaks of war, outbreaks of peace, tragic deaths of celebrities, and so on. Such events were selected because they were inferred to capture a large percentage of the world's attention.

As an example, the live television broadcast of the funeral of Pope John Paul II, on April 8, 2005, was an event that captured the devoted attention of hundreds of millions of people around the world. Before the funeral, Roger Nelson predicted that the network of RNGs would show a significant deviation from the beginning to the end of the funeral. The GCP data shifted as predicted, with odds of 42 to 1, and then returned to chance within hours of the funeral (Figure 11–5).

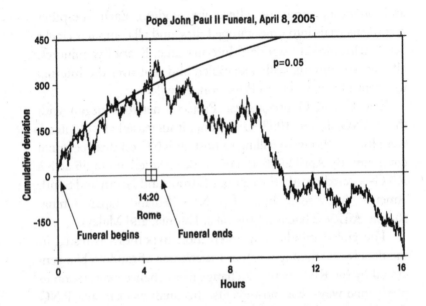

Figure 11-5. Deviation from chance for the Global Consciousness Project network of RNGs, from the beginning of the funeral of Pope John Paul II, on April 8, 2005, to 16 hours later. The parabola shows the threshold for odds against chance of 20 to 1. The significant rise in the curve during the funeral is predicted by the field-consciousness hypothesis, which assumes that the coherent mental attention of millions of people is reflected by an increase in physical order in the environment.

From August 1998 through April 2005, 185 such events were evaluated.[24] The overall results show a clear deviation from chance, with odds against chance of 36,400 to 1 (Figure 11-6). This suggests that when millions to billions of people become coherently focused that the amount of *physical* coherence or order in the world also increases. These moments of unusual coherence would not just be limited to RNGs, but would affect everything. That is, presumably every animal, plant, and rock would behave slightly differently during moments of high global coherence. We notice the effects in RNGs because we're continuously monitoring them, and we know how to spot unusual forms of order in these devices. But the hypothesis in this experiment extends to (at least) the entire globe.

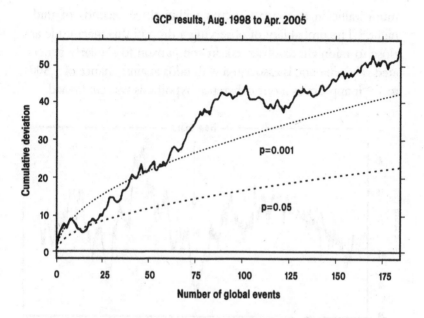

GCP results, Aug. 1998 to Apr. 2005

Figure 11-6. Results for 185 events in the Global Consciousness Project where the analyses were predefined before examining the data. The parabolas show thresholds for odds against chance of 20 to 1 and 1,000 to 1. Overall the odds against chance cumulate to 36,400 to 1, suggesting that events capturing mass global attention appear to generate moments of physical order, as detected in the outputs of RNGs located around the world.

A unique event that we studied was the stroke of midnight from 1999 to 2000, known as Y2K. Many around the world anticipated this moment with special dread and excitement. Prophecies of Armageddon combined with predictions of a worldwide computer meltdown made it a memorable New Year's Eve, and thus a good test case for this experiment. Before Y2K, I predicted that as the stroke of midnight approached in each time zone, the increasingly coherent attention of millions of people would increase the order detected by the RNGs.

After Y2K came and went, I was pleased to find that the world was still here, so I proceeded to analyze the data. The result showed that the variance or "noise" among the RNGs dropped precipitously at the stroke of midnight (Figure 11-7). The minimum value in this curve occurs within three seconds of midnight.[25] The probability of observing a drop of this magnitude as close to midnight as observed, in comparison to similarly generated random data, is associated with odds against chance of 1,300 to 1.[26] It appears as if our coherence hypothesis was confirmed.

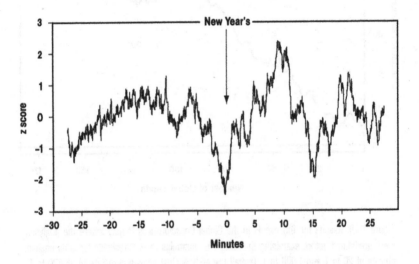

Figure 11–7. Global Consciousness Project random data for 30 minutes before and after Y2K, the transition from 1999 to 2000, based on New Year's celebrations in each time zone, around the world.[27] The dip in the curve within seconds of midnight indicates that the variance or "noise" measured among the RNG outputs plunged as people in each time zone anticipated the stroke of midnight.

One question that comes to mind when considering the Y2K effect is whether the same sorts of deviations were observed in high and low population time zones. An estimated 6 billion people live in 19 time zones over the major land masses, compared to only 9 million living in 10 time zones on islands in the Atlantic and Pacific Oceans. If the "cause" of the drop in randomness is the coherent attention of people celebrating midnight in each time zone, then we might expect that the *difference* in randomness between high and low population time zones would drop as we approach midnight. This was indeed the result, as measured across all New Year's Eve data from 1999 to 2005 (Figure 11–8). The odds against chance of the observed drop was 80 to 1. So again, these data seem to confirm our guess that mass mind "moves" matter.

At about this point I'm often asked, What about the whales and dolphins? Or the billions of fish? Or trillions of insects? Doesn't this experiment assign a bit too much importance to the interests of human beings? The answer is yes, the Global Consciousness Project is human-centric, but not because of lack of interest in other creatures. The problem is that we don't know when whale New Year's occurs, nor whether insects have special days of celebration. If we did, it would certainly be interesting to see whether fluctuations in randomness correspond to those events. Of course, we also don't know how important the "amount" of conscious awareness is in generating these outcomes, and it may turn out the coherent attention of five humans is equivalent to the attention of five trillion ants. But this remains an open question.

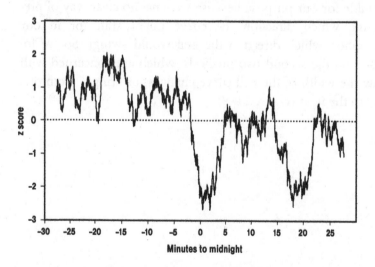

Figure 11–8. Difference in randomness in GCP data between high- and low-population time zones. The magnitude of the observed drop, within a few seconds of midnight, is associated with odds against chance of 80 to 1. This suggests that the coherent action of billions of minds in the high-population time zones was responsible for the drop in variance shown in Figure 10–7.

FOR WHOM THE BELL TOLLS

Perhaps the most dramatic event examined by the project so far occurred on September 11, 2001. On that day of infamy, now known as 9/11, we found numerous striking changes in the randomness network. To explain the nature of these anomalies and to better appreciate why the results we found are not due to any number of mundane flaws or mistakes, it's useful to think of the data that the GCP network produces as a kind of bell. That is, each RNG in the network continually generates sequences of random bits that, if sampled periodically, form a distribution resembling a bell-shaped curve. There are four simple ways that a bell curve can deviate from a theoretically perfect bell shape. It can be shifted to the left as compared to chance expectation, shifted to the right, squashed flat (the top of the bell pushed down), or squashed thin (the sides of the bell pushed towards the center). The first two deviations are not suitable for our purpose because we have no clear way of predicting which direction the curve might shift (or in our metaphor, which direction the bell would swing). So we focused on the second two methods, which are concerned with how the width of the bell curve changes, or "rings," from one day to the next (Figure 11–9).

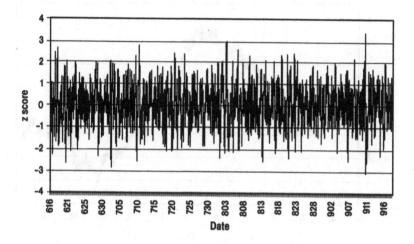

Figure 11-9. "Ringing of the bell" associated with RNG data collected from June 16, 2001 (noted as "616" on the X-axis) through September 20, 2001. In this graph, values on the Y-axis less than –2 or greater than +2 are essentially random noise, while values above and below this range occur far less frequently and are thus more interesting. Notice that on only one day these values deviated beyond +3 and –3. That day was September 11, 2001.

In examining the results of this analysis, we noticed that something unusual happened one day. On September 11, 2001, the curve deviated wildly as compared to all the other days we examined (Figure 11-10). As it happened, this curve peaked nearly two hours before a hijacked jet crashed into World Trade Tower 1 in New York City at 8:46 a.m. EDT, and it dropped to its lowest point around 2 p.m., roughly eight hours later. There's no easy answer for why the peak in this curve occurred before the terrorist attacks unfolded, although it is reminiscent of the data obtained in the presentiment experiments described in Chapter 10.[28] The huge drop in this curve within an eight-hour period was the single largest drop for any day in the year 2001. In metaphorical terms, it means the GCP bell rang loudest on that day.

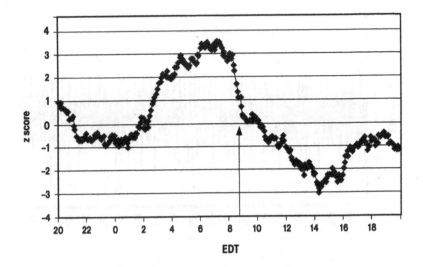

Figure 11–10. Close-up of the ringing GCP bell across 36 RNGs from 8 p.m. September 10, 2001 to 8 p.m. September 11, 2001. This was the largest change observed in the year 2001. The X-axis is in hours, Eastern Daylight Time. The arrow points to when the first hijacked jet crashed into the World Trade Center in New York City. Notice that this curve peaked a few hours before the terrorist events unfolded, possibly suggesting an anticipatory effect.

What caused this large change? Did the massive coherence of mind on that day induce a massive coherence that was reflected in the RNGs? It appears so. To further test this idea, I created a daily measure of the degree of similar behavior among the different RNGs. I called this measurement an "intercorrelation" value. Examination of this measure for every day in 2001 showed that 9/11 had the largest intercorrelation. This means that the GCP bell rang loudest on that day because all of the RNGs behaved *in the same way,* even though they were located hundreds to thousands of miles apart, scattered around the world.[29]

Could there be a mundane explanation for this effect? Could unusual environmental effects, such as increased cell-phone usage on that day, have caused this large intercorrelation effect? If that were the case, we might expect to see a few very high relationships among RNGs located in, say, North American and European cities, where cell phone usage is high, and most of the other correlations should be at chance. But this wasn't the case. The intercorrelations were distributed more or less uniformly around the world, implying that all of the RNGs "rang" in unison more than usual.

The next question of interest was whether these worldwide field-consciousness effects generalize to other days. That is, does it take a gigantic global event to catalyze an effect large enough to be observed in the randomness network, or are ordinary daily fluctuations in the world's attention also affecting randomness? To test this, I needed an objective measure of "newsworthy events." I decided to use all of the news events listed in the "Year in Review" feature on the *Information Please* Web site for the year 2001.[30] I selected this Web site over other online news sources, such as CNN, because it provides a day-by-day listing of news *events,* whereas most other sites list important news *stories,* such as "the economy," without providing day-to-day historical details.

For the one-year test period, a total of 394 news events were listed for 250 days. If the GCP network really was responding to the world's attention to global events, then I predicted that those 250 newsworthy days would have a larger average intercorrelation value than the remaining 115 non-newsworthy days. This was confirmed with odds against chance of 100 to 1. Then I tried a more general test to see if the "amount" of daily news was related to the daily RNG intercorrelation values. To create this quantity, I counted the number of letters used in the daily description of news events. This is an admittedly crude but entirely objective measure of the amount of news per day, because many news events on the same day would lead to more letters. The mind-matter interaction hypothesis predicts a positive relationship between the amount of news and the RNG intercorrelation values. Again, the result was indeed significantly positive, with odds against chance of 1,000 to 1.[31]

After discussing these analyses with three colleagues who had each been independently analyzing these data and finding similar anomalies (psychologist Roger Nelson, computer scientist Richard Shoup and physicist Peter Bancel), we jointly decided to bring these effects to the attention of the broader scientific community. So we published an article focusing on the statistical anomalies associated with 9/11 in the journal *Foundation of Physics Letters*.[32] One of those anomalies was that the GCP random data showed an extremely unlikely and persistent temporal structure called an autocorrelation, with odds against chance of a million to 1. This autocorrelation means that *something*, perhaps changes in mass attention, caused the random data to behave in a dramatically nonrandom way on 9/11, whereas it behaved normally on other days (Figure 11–11).

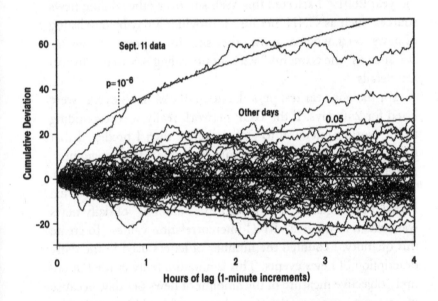

Figure 11–11. Autocorrelation (a measure of how similar data is over time) for RNG data on September 11, 2001, contrasted with the identical analysis for 60 surrounding days. The upper parabola is the threshold for odds against chance of a million to 1. This suggests that the intense mental coherence experienced on that tragic day changed the world in more ways than we knew.

Overall, the field-consciousness experiments suggest that the small mind-matter-interaction effects previously observed only in the laboratory also appear in the uncontrolled contexts of real life, possibly even at a global scale. Besides providing a new type of evidence for collective psi, these studies are exploring a new twist on the enduring riddle, "For whom does the bell toll?" Perhaps John Donne had the right answer in the sixteenth century: "No man is an island . . . therefore never send to know for whom the bell tolls; it tolls for thee."

CHAPTER 12
A NEW REALITY

All of physics is either impossible or trivial. It is impossible until you
understand it and then it becomes trivial.

—Ernest Rutherford

Now we're ready to shift gears from examining the experimental evidence to exploring ways of understanding psi. While
we're shifting, it's useful to reiterate the essential conclusion so
far: The existence of a few basic psi effects is now sufficiently
well established to persuade most scientists who study the evidence with a critical eye, and without prejudice, that something
interesting is going on. The "something interesting" here is profoundly important from a scientific perspective and deserves serious attention.

That said, it's also important to clarify that there are many
other claims associated with psi where the scientific evidence
isn't very persuasive, or where the claims could not be verified
under scrutiny, or where interpretations of the experimental results are still ambiguous. This includes large-scale physical effects like levitation and teleportation, smaller-scale physical
effects like metal bending and movement of small objects,
claims of "psychic surgery," the precise nature of apparitions
and out-of-body experiences, and so on.[1]

Also bear in mind that just because there's reason to believe that a few psi effects are real, this doesn't automatically mean that everything "paranormal" is suddenly true. Claims of Elvis and Bigfoot drag racing UFOs in the Bermuda Triangle should not be confused with the results of controlled laboratory experiments.[2] Maintaining an open mind is essential when exploring the unknown, but allowing one's brains to fall out in the process is inadvisable.

So the question then becomes, with our brains intact, and challenged by evidence for genuine psi, is there a way to understand these experiences without resorting to occultism or mythology? To answer this question it's useful to begin by reviewing how scientific assumptions about reality have changed over the last century. As shocking as it might have seemed to seventeenth-century common sense, by the twentieth century physicists had effectively revised all of the assumptions that had launched modern science three centuries before. In the twenty-first century, I believe we will continue to find increasingly strong reasons to believe that some of the strange effects observed in the microscopic world exist not only in exotic realms, but also in the more intimate domain of human experience. I also believe that the implications of all this for understanding psi are sufficiently remote from engrained ways of thinking that the first reaction will be confidence that it's wrong. The second will be horror that it might be right. The third will be boredom because it's obvious. Let's see why.

THE CLASSICAL WORLD

Classical physics began in the seventeenth century when pioneers such as Italian mathematician Galileo Galilei, French philosopher René Descartes, German astronomer Johannes Kepler, and English mathematician (and alchemist) Isaac Newton advanced a new idea. The idea was that through experiments one could learn about Nature, and with mathematics, describe and predict it. Thus rational empiricism was born. Classical physics was extended and substantially refined in the nineteenth and twentieth centuries by luminaries like James Clerk Maxwell, Albert Einstein, and hundreds of other scientists. Today phrases like "Newtonian physics" and the "Newtonian-Cartesian worldview" are used to refer to this long line of very successful inquiry.

Classical physics rests upon five basic assumptions about the fabric of reality: *reality, locality, causality, continuity,* and *determinism.* These assumptions were postulated to take place within a framework of an absolute fixed space and time. It was also taken for granted that the mathematical descriptions of physical processes corresponded to the actual behavior of objective events.

The assumption of *reality* refers to the idea that the physical world is objectively real. That means it exists independently of whether anyone is observing it. The moon is still there even if you aren't looking at it. *Locality* refers to the idea that the only way that objects can be influenced is through direct contact.[3] Unmediated action at a distance is prohibited, as this is uncomfortably close to the occult suggestion that invisible spirits can cause things to occur, and occult concepts are anathema to science.

Causality assumes that the arrow of time points only in one direction, and thus that cause→effect sequences are absolutely fixed. *Continuity* assumes that there are no discontinuous jumps in nature or that the fabric of space and time is "smooth." *Determinism* assumes that, as Einstein once quipped, "God does not play dice with the universe," meaning that things progress in an orderly, predictable way. We might not be smart enough or know enough to predict everything, but determinism says that in principle we can predict the future completely if we knew all the starting conditions and causal linkages.

CLOUDS ON THE HORIZON

Science developed rapidly with these commonsense assumptions, and classical ways of explaining how the world works are still used to explain large segments of the observable world, from particle physics to the neurosciences to cosmology. It's applicable for most objects at the human scale.

In fact, it was so sensible that the prevailing opinion among prominent scientists of the late nineteenth and early twentieth century was that physics was just about wrapped up. Physicist Albert Michelson, the first American to win the Nobel Prize, offered the following oddly pessimistic remarks in an 1894 dedication speech at the University of Chicago's Ryerson Physics Laboratory, which would later become a key player in the U.S. government's secret project to build the atomic bomb during World War II. Michelson said:

> The more important fundamental laws and facts of physical science have all been discovered, and these are now so firmly established that the possibility of their ever being supplanted in consequence of new discoveries is exceedingly remote. . . . Our future discoveries must be looked for in the sixth place of decimals.

Six years later, Scottish physicist William Thomson, president of the Royal Society from 1890 to 1895, and later granted the title of Lord Kelvin for his contributions toward creating a transatlantic telegraph cable, echoed Michelson's sentiment in a lecture presented in 1900. Kelvin said, "There is nothing new to be discovered in physics now. All that remains is more and more precise measurement."[4] In that same lecture, Kelvin spoke of two "small clouds on the horizon of physics." They both referred to properties of light.

One problem was that many physicists, including Isaac Newton, had postulated that light was composed of particles while others assumed that it was a wave. Indeed, it seemed to have properties of both. Then in 1801 British physician and physicist Thomas Young conducted experiments on light using a double-slit apparatus, which demonstrated the wave property of light so clearly that most physicists were swayed into accepting that light must be a wave. As a wave, light was assumed to display the same sort of properties as water waves, including interference and diffraction (bending around objects). Thus it was quite reasonably assumed that light, like water, was an energetic "waving" within some sort of medium. This invisible medium was called the "luminiferous ether," and it was assumed to permeate all space. Unfortunately, experiments conducted in 1887 by Albert Michelson and chemist Edward Morley failed to unambiguously detect the ether.[5]

Another problem was that when an object was heated, like a chunk of metal in a fireplace, the intensity of the light produced as the temperature was increased was predicted to be proportional to the frequencies (the colors) of light produced. The classically predicted curve worked well for low frequencies but not for higher frequencies. The problem was dubbed the "ultraviolet catastrophe" because at higher frequencies of light, the ultraviolet range and above, energies were predicted to become catastrophically immense. And that didn't happen. In 1900, German physicist Max Planck offered a solution to the problem. He developed a mathematical description of the observed light intensity that assumed light existed only in discrete energy packets. He dubbed those energy packets "quanta," and the quantum era was born. Planck was awarded the Nobel Prize in 1918.

There was another problem involving light, known as the "photoelectric effect." This refers to the observation that if you shine light on a piece of metal, electrons will be knocked loose and the flow of electrons can be detected as an electric current. But not all colors (or wave frequencies) of light did this. There was a wavelength threshold below which no electrons were released. A red light would not produce a current, and yet a blue light would. The classical idea of light as a continuous wave of energy did not account for this observation, as a very bright red light should presumably carry a lot of energy and a weak blue light should carry a little.

In 1905, an unknown Swiss patent clerk named Albert Einstein used Planck's idea of the quantum, which many scientists had regarded as little more than a mathematical trick, to successfully solve this problem. He found that light, considered as a particle with discrete energies, could account for the photoelectric effect. Light was now considered to have both particle and wave properties. Einstein was awarded the 1921 Nobel Prize for this discovery.

The quantum idea began to catch on. In 1913, Danish physicist Niels Bohr showed how the quantum concept could explain the structure of the atom (1922 Nobel Prize). In 1924, Louis de Broglie proposed that matter also has wavelike properties (1929 Nobel Prize). In 1926, Erwin Schrödinger developed a wave-equation formulation of quantum theory (1933 Nobel Prize).

PARTICLE-WAVES

All this heady progress evoked new problems. It had become increasingly clear that light had properties of both particles and waves. Particles are like billiard balls; separate objects with specific locations in space and "hard" in the sense that if hurled at each other with sufficient force they would produce the energetic equivalent of fireworks. In contrast, waves are like ripples in water. They are not localized in space but spread out, and they are soft in that they can interpenetrate and interfere with each other without harm.

This contradictory property of light was a serious problem that physicists continued to debate for many years (and still do). In 1927, Werner Heisenberg formalized the curious wave-particle relationship in his uncertainty principle: "The more precisely the position of a particle is determined, the less precisely its momentum is known, and vice versa." This uncertainty is not due to our ignorance about the position or momentum of a photon or particle but rather to a fundamental limitation of knowledge one can gain about systems that have complementary properties, like waves and particles, or positions and momentum. The uncertainty arises from the wave-like properties inherent in the quantum mechanical description of nature.

Complementarity arises in quantum physics because the mathematics of uncertainty are *noncommutative*. This means that the order of the multiplication of terms is important. Unlike the mathematics of everyday life, where $A \times B = B \times A$, in noncommutative systems $A \times B \neq B \times A$. This means that a physical system, like a photon, that consists of properties A and B, like a particle and a wave, cannot simply be decomposed into two separate subparts. So a photon cannot be considered as just a particle or just a wave. It's a *mixture* of both.

The strange consequences of the noncommutative properties of the mathematics of quantum theory was first pointed out by Einstein in a paper coauthored by Boris Podolsky and Nathan Rosen, known as the "EPR paper."[6] This paper was meant to challenge quantum theory, arguing that the theory couldn't possibly be true because surely Nature did not allow for such bizarre "mixed" properties to exist. Heisenberg's uncertainty principle also challenged basic assumptions about the nature of causality. According to the classical understanding of causality, if we know the present state of a particle exactly, then we can exactly calculate its future state. But the uncertainty principle says that we cannot know *all* of the present properties of a particle, and thus the future cannot be determined, even in principle. This was a radical idea that physicists refused to accept for many decades, but it did fit within the framework of quantum theory.

THE ONLY MYSTERY

In 1990, Richard Feynman, one of the more colorful Nobel laureates of the twentieth century, said:

> What I am going to tell you about is what we teach our physics students in the third or fourth year of graduate school. . . . It is my task to convince you not to turn away because you don't understand it. You see my physics students don't understand it. . . . That is because I don't understand it. Nobody does.[7]

Feynman was referring to a key experiment demonstrating the complementarity of the atomic world: Thomas Young's double-slit experiment, first described in 1801. Two centuries later, readers of *Physics World* magazine voted that experiment as "the most beautiful experiment" in physics.[8] In describing this experiment, Feynman said "We choose to examine a phenomenon which is impossible, *absolutely* impossible, to explain in any classical way, and which has in it the heart of quantum mechanics. In reality, it contains the only mystery."

In the classic double-slit experiment, a stream of photons (or electrons or any atomic-sized object) are shot at a screen with two tiny slits in it. On the other side of the screen, a photographic plate or sensitive video camera records where each photon lands (Figure 12–1). If one of the slits is closed, then the camera will see a smooth distribution of photons with the peak intensity directly opposite the open slit. This is what common sense would predict if the photons were individual particles. But if you open both slits, the camera sees a different pattern: an interference pattern with varying bands of high and low intensity. That is consistent with the photons being a wave.

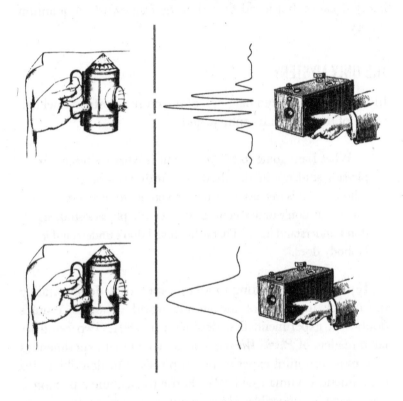

Figure 12–1. Double-slit experiment with two slits and one slit open.

What happens if you lower the light intensity so only one photon is sent through the apparatus at a time? You'd expect with only one slit open to see a single, smooth distribution as before, and indeed that is what you find. But you might also expect that with two slits open you'd see two smooth distributions, one behind each slit. After all, individual photons are shot at the screen, so each photon must presumably go through just one of the slits. But this is not what happens. If you shoot individual photons through the apparatus, one at a time, after collecting lots of them you'll eventually end up with same interference pattern that you saw when you shot a flood of photons through the screen.

This means that each photon individually goes through *both slits at the same time,* as though each photon behaved as a wave. In that state it interfered with or was *entangled with itself.* It's important to clarify that we never actually observe a photon (or electron, or any other atomic object) as a wave. It just acts that way when we're not looking!

To make things more perplexing, imagine that you have an extremely fast shutter in front of one of the slits, so you can open or close it *after* the photon has gone through the slit but *before* it has been captured by the camera. Now you fire a single photon at the apparatus and measure where it lands. Sometimes the slit is open and sometimes closed. What you find is that even though the decision to open or close the slit was made *after* the photon had already gone through one or both slits, the resulting behavior is like a particle if one slit is closed and like a wave if both slits remain open. Somehow the photon "knows" *after* it had already gone past the slits that one of them would be closed *later.*[9] Physicist John Wheeler proposed this experiment and dubbed it a "delayed-choice" design. When tested in the laboratory, this is indeed what happens. This means that the photon is entangled with itself not only in space, but also in *time.*

Now to make things truly mind-bending, there is something called the "delayed-choice quantum eraser." This effect, proposed in 1982 by physicists Marlan O. Scully and Kai Drühl indicates that the "erasure of information generated in the past can affect how we interpret data in the present."[10] In other words, after the experiment is finished and the photon's position is already recorded, the experimenter can still decide whether that photon had passed through one slit or both slits![11] In thinking about these curious results, it's wise to heed Feynman's advice:

Do not keep saying to yourself, if you can possibly avoid it, "But how can it be like that?", because you will get "down the drain" into a blind alley from which nobody has yet escaped. Nobody knows how it can be like that.

Not only does no one know how it can be like that, but this rather simple experiment raises a central question about the role of the observer in quantum reality. This is known as the *quantum measurement problem*: We infer that the photon acts like a wave when we're not looking, but we never actually *see* those waves. So what causes the photon to "collapse" into a particle when we do decide to look at it?

In classical physics, objects are regarded as objectively real and independent of the observer. In the quantum world, this is no longer the case. As physicists Bruce Rosenblum and Fred Kuttner from the University of California at Santa Cruz put it, "If we assume that no observable physical phenomena exist other than those specified by the present quantum theory, a role for the observer in the experiment can be denied only at the expense of challenging the belief that the observer makes free choices."[12]

In other words, the experimenter's *choice* of whether to keep one of the slits open or closed changes how the photon behaves.[13] This effect of our choice does not depend on the objects used. Any quantum object would show the same results, and given that all physical objects are already quantum objects, this is a general question not just limited to the microscopic world. Successfully *detecting* how larger objects change as a result of our observation is limited by how technically sophisticated we may be, but there is no fundamental limitation. In other words, according to Rosenblum and Kuttner, an experimenter must conclude that:

Reality was somehow created by the observation *itself*, that the observed reality is created *solely* by the observer's acquisition of knowledge. If so, the observer is inseparably involved with the observed system. That would challenge his view of a physically real world existing independently of his senses perceiving it. The only alternative the experimenter sees to this observer-involved reality is to question his ability to freely choose the experiment.

Few of us believe that everything we do is predetermined, or that we have absolutely no free will. Certainly we must *act* as though we have free will, in fact the legal system insists upon it. Otherwise you run the risk of being declared incompetent, and then you'll spend the rest of your life enjoying a vacation in a mental hospital.

But even if it were conceded, as some neuroscientists insist, that free will is an illusion, it still leaves open the undeniable relationship between what *we* personally decide to do and what the photon ends up doing. So what causes that relationship? This question remains unsolved. After reviewing this problem, Columbia University physicist Brian Greene concluded: "After more than seven decades, no one understands how or even whether the collapse of a probability wave really happens."[14] Many physicists, like Rosenblum and Kuttner, question the role of the mind in *literally* affecting the physical world:

> The measurement problem arising from the quantum experiment does not necessarily imply that something "from the mind of the observer" affects the external physical world. The measurement problem does, however, hint that there is more to say about the physical world than quantum theory says.

But others, like Pascual Jordan, one of the principal mathematical architects of quantum theory, wasn't so sure. In his view:

>   Observations not only disturb what has to be measured, they produce it . . . . We compel [the electron] to assume a definite position . . . . We ourselves produce the results of measurement.[15]

As we've seen, the evidence from psi experiments suggests that observation, in the form of attention and intention, does seem to influence the world. Thus Jordan may have been right after all.

CHANGING ASSUMPTIONS

The mysteries of quantum theory revolve around the concepts of superposition, complementarity, uncertainty, the measurement problem, and entanglement.[16] All of these concepts are, in effect, different ways of pointing at the same puzzle: In the un- observed state, a quantum object does not have definite location in time or space, nor does it have definite properties, at least not in the way that we think of *definite* in classical terms. How can something be said to exist if it doesn't have proper- ties, location, or existence in time? We don't know, but it sug- gests that something about our ordinary assumptions of an objective classical reality "out there," independent of us, is mis- taken.[17]

Perhaps the most startling consequence of this "new reality" is that the assumptions of classical physics, and common sense, have significantly softened from their previous absolute status. An absolute *reality* independent of us fades like the Cheshire Cat because we now know that fundamental properties of the world are not determined before they are observed. This is not to say that reality doesn't exist, but rather that an unobserved reality is radically different than the one we're familiar with. In the EPR paper, Einstein questioned if quantum theory could possibly be correct because among other things, it implied that the moon wasn't there if you weren't looking at it. Cornell University physicist N. David Mermin considered Einstein's complaint in light of the experimental data, and concluded, "Einstein maintained that quantum metaphysics entails spooky actions at a distance [*spukhafte Fernwirkungen*]; experiments have now shown that what bothered Einstein is not a debatable point but the observed behavior of the real world."[18]

The new reality has replaced the assumption of *locality* with the concept of *nonlocality*. The fact that quantum objects can become entangled means that the common sense assumption that ordinary objects are entirely and absolutely separate is incorrect. In unobserved states, quantum objects are connected instantaneously through space and time. It is no longer the case that unmediated "action at a distance" is prohibited because it's spooky. In fact, unmediated action at a distance in quantum reality is *required.*

The new reality has dissolved *causality* because the theory of relativity revealed that the fixed arrow of time is an illusion, a misapprehension sustained by the classical assumptions of an absolute space and time. We now know that *when* events seem to occur depends on the perspective (technically, the frame of reference) of the observers.

The new reality has abandoned the assumption of *continuity* because the fabric of quantum reality is discontinuous; at small scales, space and time are neither smooth nor contiguous. And finally, absolute *determinism* has been fatally challenged because it relies on the assumptions of causality, reality, and certainty, none of which exist in absolute terms anymore.

Dissolution of the classical assumptions has also challenged a very basic approach to how science understands the world, an approach known as *reductionism*. Scientists have long assumed that the best way, indeed perhaps the only way, to understand something is see how its pieces fit together. If we see an impressive clock and want to know what makes it tick, we take it apart. In medical research the entire focus is on understanding the "mechanism of action" of a treatment. All of this assumes that "how things work" involve mechanical processes, like interlocking gears.

Many processes certainly *appear* to be explainable in approximately mechanistic, reductionistic terms. But as physicists have delved progressively deeper into the nature of reality, they find that it cannot be understood in mechanistic terms. Mechanism assumes that there are separate objects that interact in determined, causal ways. But that's not the reality we live in. Quantum reality is holistic, and as such any attempt to study its individual pieces will give an incomplete picture. It's like studying atoms inside an acorn in an attempt to understand the emergence of leaves on an oak tree—a futile exercise.

Few physicists today doubt that quantum theory provides an accurate description of the observable world. For example, its prediction of the strength of interactions between an electron and a magnetic field has been experimentally confirmed to a precision of two parts in a trillion. For a theory that is so preposterously precise, it's equally preposterous that there is no widespread agreement yet over what it *means*.

COPENHAGEN INTERPRETATION

One of the leading interpretations of quantum theory was advanced principally by Niels Bohr of the Institute for Theoretical Physics at the University of Copenhagen, now known as the Niels Bohr Institute. The Copenhagen interpretation is orthodox in the sense that it was the first widely accepted interpretation, and it's the one still favored today by most physicists. It says that ultimately quantum theory tells us what we can *know* about reality rather than about reality itself. This may be regarded as a "don't ask, don't tell" sort of interpretation, which allows the theory to be used without having to worry about what it means. Bohr employed this strategy to avoid having to grapple with the possibly unanswerable questions it posed.

Bohr's approach introduced a major change over classical

physical assumptions in that now it was no longer possible to assume that we, the experimenters and observers of physics experiments, were separate from the experiments themselves. As he put it, "In the great drama of existence we ourselves are both actors and spectators."[19]

MANY WORLDS

Proposed by physicist Hugh Everett, the many worlds interpretation suggests that when a quantum measurement is performed, every possible outcome actually manifests. This avoids the problem associated with the role of the observer, in that here we assume that the observer isn't important at all. Instead, in the process of a quantum possibility manifesting into an actuality, the universe splits into two, or as many versions of itself as needed to accommodate all possible measurement outcomes. Popular television shows like *Sliders, Quantum Leap,* and *Star Trek* have capitalized on this idea of multiple, parallel universes for the interesting story line it suggests: Untold trillions of new universes are being created each instant, each similar to the others except that after splitting they enjoy their own, separate evolutions. Many physicists are uncomfortable with this idea because it colossally violates the principle of parsimony, the preference in science for the simplest possible explanation.

QUANTUM LOGIC

Another interpretation proposes that we are confused by the implications of quantum theory because commonsense assumptions about logic no longer hold when dealing with complementary systems. Quantum logic demands that a photon be either a wave or a particle, but not both. Or that a number be either 0 or 1, but not both 0 and 1 at the same time. But because experiments show that photons do act this way, it seems that the either-or logic of common sense no longer holds in the quantum world. Until our language and logic evolve to more easily grasp complementary ideas, it's likely that we'll continue to experience confusion and paradoxes.

CONSCIOUSNESS CREATES REALITY

Still another interpretation proposes that the act of observation literally creates physical reality. In its strong form, this interpretation asserts that consciousness is the fundamental ground state, more primary than matter or energy. This position provides a special role for observation by becoming the active agent that collapses quantum possibilities into actualities. Many physicists are suspicious of this interpretation because it resembles ideas originating from Eastern philosophy and mystical lore. But a notable subset of prominent physicists, including Nobel laureate physicists Eugene Wigner and Brian Josephson, John Wheeler, and John von Neumann have embraced concepts that are at least mildly sympathetic to this view. Physicist Amit Goswami, from the University of Oregon, has strongly promoted this view.[20]

DECOHERENCE

One interpretation attempts to account for the difference between the nonlocal "virtual" reality assumed to exist in the unobserved quantum state versus the local realism we always experience in the world of actual observations. The decoherence interpretation rests upon the Copenhagen interpretation but delves deeper into the question of what happens at the boundary between the observed and unobserved. It assumes that when a quantum object interacts with the environment, those interactions act as "observations" and as such rapidly smooth out (decohere) the quantum discreteness and "collapse" it into classical-looking behavior. However, as physicist Brian Greene notes:

Even though decoherence suppresses quantum interference and thereby coaxes weird quantum probabilities to be like their familiar classical counterparts, *each of the potential outcomes embodied in a wavefunction still vies for realization.*[21]

This failure to cleanly resolve the measurement problem led University of Maryland Philosopher Jeffrey Bub to refer to decoherence theory as an "ignorance interpretation."[22] British mathematician Chris Clarke, of the University of Southampton, further points out that

Decoherence is the losing of quantum information to the environment; but the universe as a whole *has no environment.* Cosmologically, information is never lost . . . . This suggests . . . that the universe remains coherent; it was, is and will always be a pure quantum system. The non-coherence of medium scale physics—non-coherence "for all practical purposes," as John Bell used to say—is only an approximate consequence of our worm's-eye view.[23]

NEOREALISM

Einstein, who could not accept the outlandish implications of quantum theory, favored the neorealism interpretation. This is the interpretation assumed by the majority of scientists who don't spend much time thinking about quantum theory. Neorealism proposes that reality consists of objects familiar to classical physics, and that the strangeness of quantum theory can be accounted for by our ignorance of hidden variables. Once those additional factors are discovered, it is assumed that the quantum weirdness will be completely understood, and then local realism and common sense will once again reign supreme. Einstein's 1935 "EPR paper," entitled "Can quantum mechanical description of physical reality be considered complete," provided an argument against the nonlocal, nonrealistic view suggested by quantum theory.[24] Einstein protested: "I cannot seriously believe in [the quantum theory] because it cannot be reconciled with the idea that physics should represent a reality in time and space, free from spooky actions at a distance."[25] It was in a discussion of the EPR paper that Erwin Schrödinger first coined the term "entanglement." He wrote:

> If two separated bodies, each by itself known maximally, enter a situation in which they influence each other, and separate again, then there occurs regularly that which I have just called *entanglement* of our knowledge of the two bodies[26]. . . . I would not call that *one* but rather *the* characteristic trait of quantum mechanics.[27]

For decades, the argument over possible hidden variables that might reestablish ordinary reality revolved primarily around one's philosophical preferences. But in 1964, Irish physicist John Bell mathematically proved that no local hidden variables theory could be compatible with quantum theory. "Bell's theorem" has subsequently been described as "the most profound discovery in science."[28] Before Bell's theorem, all experimental tests had confirmed quantum theory's predictions, so most physicists considered it unlikely that Einstein's complaint was correct. And yet, at the same time most believed, like Einstein, that at its core the world really was locally realistic—the moon really is there when you don't look at it. Unfortunately, unless reality itself is complementary, and it can be both local and nonlocal at the same time, one or the other belief *must* be false.

In 1964 the debate swung strongly in favor of quantum theory through Bell's theorem, and that led to a series of increasingly persuasive experimental tests starting in 1972.[29] Perhaps the best known of these experiments were those by French physicist Alain Aspect and his colleagues at the Institut d'Optique in Orsay, France, in 1982, and in 1998, an even more conclusive series of tests by Nicholas Gisin's group at the University of Geneva. In the latter case, nonlocal entanglement of photons was demonstrated over 11 km of optical fiber. In 2004 Gisin's group repeated the result over 50 km.[30]

The repeated confirmation of quantum theory, and of the concept of entanglement, would have genuinely shocked Einstein. As physicist Daniel Greenberger put it, "Einstein said that if quantum mechanics was correct then the world would be crazy. Einstein was right—the world is crazy."[31] Other physicists now agree. Abner Shimony and John Clauser wrote, "The conclusions from Bell's theorem are philosophically startling; either one must totally abandon the realistic philosophy of most working scientists or dramatically revise our concept of space-time."[32]

BELL'S THEOREM

So what is Bell's theorem? A non-technical way to understand it was provided by physicist N. David Mermin.[33] Conceptually it involves an experimental apparatus with three boxes (Figure 12–2). All the center box does is emit entangled pairs of objects. They could be photons, electrons, billiards, or humans. Atomic-sized objects are used in practice because entanglement is easier to produce and detect in the realm of the very small, but in principle any physical object can be used. The two side boxes, one to the left of the center box and the other to the right, are detectors used to monitor and record properties of the objects. The two detectors are completely isolated from each other. They might be 31 miles apart as in Gisin's 2004 experiment, or 31 light-years.[34]

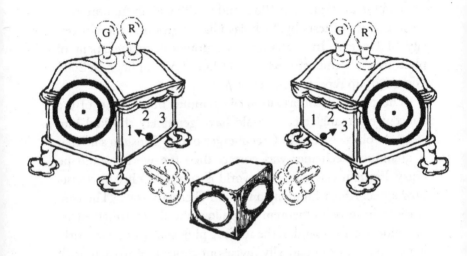

Figure 12–2. Conceptual design for understanding Bell's theorem.

Each detector box has a knob that can be placed in one of three settings, let's call them 1, 2, and 3. Each box also has two lamps, red and green; one of those lamps lights up when it detects an object. One trial in this experiment works by first selecting one of the three settings on each detector box, then pressing a button to send the entangled objects from the middle box toward the detectors, and then finally recording which of the two lights on each box lights up. In 10 successive trials of this experiment, we might end up with the following sequence of recordings:

31 RG . . . 12 GR . . . 22 RR . . . 11 GG . . . 12 RG . . .
12 GR . . . 23 GG . . . 23 GG . . . 33 RR . . . 23 GR

The numbers represent the positions of the knob settings on the two detectors and the letters indicate the lamps that lit up. Thus on the first trial, we set the left detector's selector to 3 and the right detector's selection to 1. We fire the entangled object, and then we record that a red lamp flashed on the left and a green lamp flashed on the right. It doesn't matter for this discussion exactly how the detector boxes determine which lamps to flash. What does matter is that after running lots of such trials, we find that whenever the selectors are set to the *same number* (11, 22, or 33), the lamps flash the *same color* (RR or GG). And whenever the selectors are set to different numbers, the lamps flash randomly (RR, RG, GR, or GG). Overall, we find that the sequence of lamp flashes on each individual detector box is distributed at random.

That's all there is to it: There's a correlation between the selector setting and the lamps that light up. This is the most profound discovery in science. What's the big deal?

Well, why did the detector lamps flash the same color when the knobs are set to the same selector positions? Remember, these detectors aren't allowed to communicate with each other in any way, and they haven't been preprogrammed with simple rules like "if the selector is set to 1, always flash green." When the selector numbers are the same, the lamps flash the same color, but not necessarily just green or just red.

One explanation is that since each pair of objects came from the same source, perhaps they share a similar code, like a set of tattoos that identify which gang the objects belong to, and these codes define which light will flash given for each possible selector setting. For example, say that each object shares a code that says when the selector on either box is set to position 1 the red lamp will flash, if set to position 2 the green lamp will flash, and if set to position 3 the red lamp will flash. We might call this code "Red-1 Green-2 Red-3," or RGR for short. Since each pair of objects always shares the same code, this guarantees that if the selectors are set to the same number that the lamps will always flash the same color without having to prespecify that color. There are two colors and three selector positions in our apparatus, so we have eight possible code sets: RRR, RRG, RGR, RGG, GRR, GRG, GGR, and GGG.

This explanation, involving shared codes is, as David Mermin says, the only one that "someone not steeped in quantum mechanics will ever be able to come up with." And yet, this apparently reasonable explanation simply doesn't work. Why not?

Let's assume that both objects share the code RRG (shorthand for Red-1, Red-2, Green-3). This code means that if the selector on the left box is set to 1, and the selector on the right box is also set to 1, then both boxes will flash red. If the selectors are set to 2 and 2, they'll also both flash red. And if set to 3 and 3, they'll both flash green. If we set the selectors to 1 and 2, the lamps will both flash red, and if we set the selectors to 2 and 1, the lamps will still both flash red. They'll flash different colors for the four remaining settings: 1–3 (red and green), 3–1 (green and red), 2–3 (red and green) and 3–2 (green and red). The same type of logic holds for the instruction sets RGR, GRR, GGR, GRG, and RGG, as in each of these cases five settings will cause the *same* lamps to flash, and four will cause *different* lamps to flash.

This means that five out of nine selector settings will flash the same colors for six of the eight possible instruction sets. The only two remaining instruction sets are RRR and GGG, both of which guarantee that the lamps will always flash the same regardless of the selector settings. So this means that overall the same colors will flash *at least* 5/9 or 55.5% of the time. So what? Because the actual experimental data show, and quantum theory predicts, that the lamps will flash the same colors only 50% of the time. This difference, between 55.5% and 50%, is known as *Bell's inequality*.

The concept of Bell's inequality is actually quite simple, but the first time you encounter the argument it's hard to grasp. (You may need to reread the last few paragraphs.) You'll know you've got it when your gut suddenly drops, like the feeling of freefall when a roller-coaster plunges off that first steep rise. Until you get it viscerally, the "most profound discovery" description seems like overkill. Afterwards, *profound* isn't strong enough.

You can tell when someone hasn't quite gotten this yet, because they'll patiently listen to this description, and then they'll respond that a 5.5% difference seems awfully small to have generated all this excitement. Maybe it's just due to erratic detectors that occasionally misfire, lowering the expected rate from 55.5% to 50%. For many years this possible loophole allowed people to coolly ignore Bell's inequality. But starting in 1972 the experimental evidence became progressively stronger in showing that Bell's inequality was indeed being violated. The predicted 55.5% result just wasn't showing up. In 2004, the experiments by Gisin's group violated Bell's inequality by more than 20 standard errors, a stupendously persuasive result in complete accordance with quantum theory but radically different from the predictions of classical physics theory.[35]

So the chances that this inequality is due to poor detectors is effectively zero, and this in turn means that quantum theory is correct. This is why physicists are either thrilled or disturbed (or sometimes both) by Bell's theorem. It tells us that something is unquestionably wrong about our long-held, cherished assumptions about reality. The experimental evidence has now convinced the majority of physicists that Einstein was wrong. As Brian Greene puts it,

> The most straightforward reading of the data is that Einstein was wrong and there can be strange, weird, and "spooky" quantum connections between things over here and things over there[36]... This is an earth-shattering result. This is the kind of result that should take your breath away.[37]

WHAT DOES THIS HAVE TO DO WITH PSI?

Quantum theory and a vast body of supporting experiments tell us that *something unaccounted for is connecting otherwise isolated objects.* And this is precisely what psi experiences and experiments are telling us. The parallels are so striking that it suggests that psi is—literally—the human experience of quantum interconnectedness. This may seem like an unwarranted leap, so let's ease into this idea more cautiously.

In 1909, the Harvard University psychologist William James wrote the following:

> For twenty-five years I have been in touch with the literature of psychical research, and have had acquaintance with numerous "researchers." I have also spent a good many hours . . . in witnessing (or trying to witness) phenomena. Yet I am theoretically no "further" than I was at the beginning; and I confess that at times I have been tempted to believe that the Creator has eternally intended this department of nature to remain *baffling*, to prompt our curiosities and hopes and suspicions all in equal measure, so that, although ghosts and clairvoyances, and raps and messages from spirits, are always seeming to exist and can never be fully explained away, they also can never be susceptible of full corroboration.[38]

In 1922, Columbia University psychologist Gardner Murphy, who would later become President of the American Psychological Association as well as President of the American Society for Psychical Research, received support from the Richard Hodgson Psychical Research Fund at Harvard University. Regarding his early foray into psi research, he wrote: "I thought that getting telepathy under experimental control would be relatively easy: I thought the scattered, incoherent nature of the data was due to the amateurish, puerile, half-baked amateurism of the whole business."[39] Murphy later realized that things weren't so simple.

In 2001, nearly a century after James's remarks, and after a quarter-century of their own intensive study, Princeton University researchers Robert Jahn and Brenda Dunne echoed a similar conclusion: "At the end of the day, we are confronted with an archive of irregular, irrational, yet indismissable data that testifies, almost impishly, to our enduring lack of comprehension of the basic nature of these phenomena."[40]

These are recurring themes. Scientists who realize that there's something interesting about psi are initially highly enthusiastic. Real psi is profoundly important from a scientific point of view, and at first glance it doesn't seem to be so complicated. Most psi experiments are ridiculously simple, at least in principle. So new investigators often maintain an unspoken belief that previous researchers were insufficiently competent or clever to crack the problem. Then, after a few decades of chipping away at the puzzle, the investigators are a little older and wiser and they offer a more moderated opinion. They admit that they're convinced psi really does exist, but understanding these phenomena remains an enigma.

My sense of this puzzle is that William James was on the right track. In 1897 he wrote:

> In psychology, physiology and medicine, whenever a debate between the mystics and the scientifics has been once and for all decided, it is the mystics who have usually proved to be right about the facts, while the scientifics had the better of it in respect to the theories.[41]

In this case the facts, as presented in the previous chapters, have been resolving into increasingly sharp focus over the last few decades. While some psilike experiences may be due to a combination of one or more prosaic errors, that explanation is not adequate to explain the experimental results. So on this score the mystics were probably correct about the facts. But their explanations tend to be nontestable metaphors. So far the "scientifics" haven't done much better in terms of explanations, but I believe this is going to change.

In 1909, in the same article where James confessed that psi was remarkably baffling, he also made it clear that he was convinced of the existence of real "supernormal knowledge."[42] He then offered the following metaphorical explanation:

> We with our lives are like islands in the sea, or like trees in the forest. The maple and the pine may whisper to each other with their leaves, and Conanicut and Newport [Islands near the New England coastline] hear each other's foghorns. But the trees also commingle their roots in the darkness underground, and the islands also hang together through the ocean's bottom, just so there is a continuum of cosmic consciousness, against which our individuality builds but accidental fences, and into which our several minds plunge as into a mother-sea or reservoir.
>
> Our "normal" consciousness is circumscribed for adaptation to our external earthly environment, but the fence is weak in spots, and fitful influences from beyond leak in, showing the otherwise unverifiable common connection. Not only psychic research, but metaphysical philosophy, and speculative biology are led in their own ways to look with favor on some such "panpsychic" view of the universe as this.
>
> Assuming this common reservoir of consciousness to exist, this bank upon which we all draw . . . the question is, What is its own structure? What is its inner topography? . . . Are there subtler forms of matter which upon occasion may enter into functional connection with the individuations in the psychic sea, and then, and then only, show themselves?—so that our ordinary human experience, on its material as well as on its mental side, would appear to be only an extract from the larger psychophysical world?[43]

James's cosmic consciousness metaphor has reverberated throughout the ages, ranging from ancient concepts like the Akashic record of Hindu mysticism, to psychiatrist Carl Jung's collective unconscious, to biologist Rupert Sheldrake's morphogenetic fields. But these alluring metaphors all present an unanswered question: What is the nature of this hypothetical medium in which mind and matter are intimately intertwined? Is there any independent evidence that it exists? If we see a television set and we want to know how it works, a perfectly valid explanation is, "You push this button and pictures show up on the screen." That may be literally correct, but it wasn't exactly what we had in mind when we asked, "How does that work?"

I believe that a rationally satisfying explanation for that medium is emerging, an explanation borne from physics. For it is within physics that the principal puzzle of psi resides. If physics prohibits information from transcending the ordinary boundaries of space and time, then from a scientific point of view psi is simply impossible. But here's where things become interesting. As we've seen, the old prohibitions are no longer true. Over the past century, most of the fundamental assumptions about the fabric of physical reality have been revised in the direction predicted by genuine psi.

This is why I propose that psi is the human experience of the entangled universe.[44] Quantum entanglement as presently understood in elementary atomic systems is, by itself, insufficient to explain psi. But the ontological parallels implied by entanglement and psi are so compelling that I believe they'd be foolish to ignore. It's useful to ponder that if all we knew about was the behavior of atoms, nothing would suggest that living organisms would emerge when you put those atoms together in certain ways. Even less likely would we predict the emergence of the complex structure we call conscious awareness, and still less likely could we predict a global civilization. So it's exceedingly difficult to imagine what wonders may emerge when quantum entanglement meets life. In this regard, I agree with physicist Nick Herbert, who said:

I am so amazed by the subtlety that Nature employs to operate ordinary "dead" matter that I can hardly imagine the subtlety she must deploy to operate "conscious matter." I think that our learning to understand quantum theory is kindergarten classes compared to what we will have to grok to comprehend consciousness. Not that it will be some sort of complicated mathematics but some new way of thinking.[45]

A NEW WAY OF THINKING

Now we enter a realm where normally sedate physicists use words like "outrageous," "astonishing," and "mind-boggling" to express their amazement. Let's ease into this domain by considering something simple, like a telepathy experiment.

Jack and Jill agree to participate in a new experiment.[46] When they arrive at the lab, they are taken to separate, heavily shielded chambers to prevent any form of ordinary communication of passing between them. When Jack is ready to begin sending a message to Jill, he presses a button that causes a computer outside his chamber to randomly select one picture from a set of three pictures. The computer displays the selected picture to Jack. At the same time, another computer outside Jill's chamber detects that Jack has initiated his trial, so it too randomly selects and then displays one picture to Jill based on the same picture set used for Jack.

Now Jack and Jill are asked to mentally communicate and mutually decide whether they are seeing the same or a different picture. They can answer "yes," it's the same picture, or "no," it isn't. They record their impressions in their computers. Now they try another trial, with pictures randomly selected from a new set of three pictures. Sometimes Jack initiates the trial and sometimes Jill. The data from each trial are one of four possibilities: yes-yes, yes-no, no-yes, or no-no.

They continue this test for several days until they collect a total of 1,000 such trials. Then an analysis of all the data reveals that Jack and Jill each responded "yes" about half the time and "no" half the time. But remarkably, whenever Jack and Jill viewed the *same* picture they gave the *same* answer 77% of the time instead of the chance-expected 50%.[47] That same answer might have been "yes" or "no." By contrast, when viewing mismatched pictures, they answered "yes" or "no" randomly and achieved a 50% hit rate.

By assigning a "hit" to each trial where the responses were the same, and "miss" to trials with mismatches, we find that the overall hit rate is 59%. That's the same result observed in the dream psi tests discussed earlier. With 1,000 trials, this hit rate is associated with odds against chance of 225 million to one. In short, the experiment is a great success.

Except for one minor point—this wasn't a telepathy test at all. It's an example of what quantum theory predicts if Jack and Jill were entangled. Quantum information scientist Gilles Brassard, one of the inventors of quantum cryptography, has dubbed this a *pseudo-telepathy* game in recognition of its similarity to real telepathy. In such games, Brassard says, "two or more quantum players can accomplish a distributed task with *no* need for communication whatsoever, which would be an impossible feat for classical players."[48]

At first blush, this is difficult to believe. How can two (or more) people successfully perform a joint task that requires some sort of communication, but without passing *any* signals at all? Physicist Guy Vandegrift of the University of Texas at El Paso worried about this, too. After studying the problem, he published an essay in *The Philosophical Quarterly* in which he expressed his disquiet.[49] Vandegrift began by noting that "What I recently learned . . . was so disturbing that I felt compelled to express the concept as simply as possible, without destroying the correctness of the argument. It appears that elementary particles act as if their behavior were linked by channels of communication that can be best described as 'psychic.'"

He then proceeds to describe an experiment similar to the Jack and Jill experiment just mentioned. After concluding that the results were impossible by any classical theory, and yet in accordance with both quantum theory and experimental tests that have confirmed quantum theory, he concluded that

> There seems to be no fundamental reason why two people could not put themselves into [an entangled state] and reproduce what [Jack and Jill] have done here. I did not intend to write an essay on psychic phenomena, and made this analogy because it is the most direct description of what the EPR experiment is actually doing. I do not believe in mental telepathy, miracles or any other occult phenomenon. This affair with Bell's theorem has shaken me to the bone.[50]

"Shaken to the bone" is a good way to describe how entanglement is influencing the scientific understanding of reality. It defies three centuries of scientific assumptions. And it looks and feels so much like magic that scientists who haven't thought much about entanglement are either horrified, or they deny there's any problem and vehemently refuse to explain why.[51]

Applying the concept of entanglement to disciplines beyond physics is still in its infancy, but the prospects are promising and progress has been extremely rapid. As Gilles Brassard says,

> Information theory and computer science are . . . firmly rooted in classical physics, which is at best an approximation of the quantum world in which we live. This has prevented us from tapping the full potential of nature for information processing purposes. Classical and quantum information can be harnessed together to accomplish feats that neither could achieve alone . . . .
>
> Quantum entanglement, which is the most non-classical of all quantum resources, can be used to teleport quantum information from one place to another. It enables the accomplishment of distributed tasks with a vastly reduced communication cost. In extreme cases, we can provide inputs to non-communicating participants and have them produce

outputs that exhibit classically impossible correlations: This is the mysterious realm of pseudo-telepathy.[52]

Some may regard all the excitement about entanglement a fad, or as mere hyperbole designed to annoy physicists and beguile new agers. But it goes deeper than that. Experiments have demonstrated that the worldview implied by classical physics is wrong. Not just slightly incorrect in minor ways, but fundamentally wrong in just the right way to support the reality of psi.

CHAPTER 13
THEORiES OF PSi

Albert Einstein, reporting a conversation with an unnamed but
"important theoretical physicist," regarding telepathy:[1]
He: I am inclined to believe in telepathy.
Einstein: This has probably more to do with physics than with
psychology.
He: Yes.

Psi phenomena present three key problems for theory development. The first is that information has to reach across space and time in ways that defy common sense. As Einstein said, this is a problem for *physics*. The second is that this information must arrive in your mind without the use of the ordinary senses, and it must be able to interact with objects at a distance. This is a problem for both physics and the *neurosciences*. And the third is that information must reach conscious awareness often enough for people to report it. This is a problem for *psychology* and the neurosciences.

Physics is first on the list because all three problems are closely related to our conception of physical reality. If the physical medium in which we live prohibits information from flowing in the required ways, then the only rational conclusion is that reports of psi must be due to error. From that perspective, ESP must mean "Error Some Place," despite what any experimental evidence might suggest.

Fortunately, knowledge about the nature of physical reality has evolved over the past few millennia, as we discussed in the last chapter. The direction of that evolution suggests that the "error hypothesis" is becoming less likely and the psi hypothesis more likely. To help ground this discussion, it's useful to see how concepts about the relationships among matter, mind and psi have evolved.

There are four periods in this evolution corresponding to ancient, classical, modern, and possible future eras (Figure 13-1). In ancient times the principal concept of reality was *anima mundi,* the living universe. This "Age of Magic" lasted for tens of thousands of years, and the metaphor of the times–the zeitgeist–was *spirit.* Reality was imagined to exist in cycles. This was inferred from the rhythms observed in the stars, the seasons, the days, and the lifecycles of all living creatures. It was taken for granted that the nature of reality rested upon the whims of the gods, and it was assumed that animate spirits could cause things to happen at a distance without mediation, at least without known cause within "our" world. The concept of mind was associated with the *soul,* a spark of divine presence within us. Psi phenomena in the ancient era were considered self-evident, as a natural form of communion between soul and spirits.

The next step was the classical scientific era. In classical times the overriding concept of reality was as a *mechanical universe.* This "Age of Industry" lasted from about the seventeenth century to the mid-twentieth century, and the zeitgeist was a *clockwork.* Basic physical concepts like time, space, energy, and matter were imagined to be fixed, absolute, and fundamentally different substances. It was taken for granted that reality existed in an absolute sense, independent of observers, and it was an additional token of faith that action at a distance was impossible. The concept of mind, then viewed through the fledgling discipline of psychology, and especially its rising fad of behaviorism, was regarded as an illusion created by the clockwork mechanisms of the brain. Because mind was an illusion and action at a distance was impossible, genuine psi phenomena were also impossible.

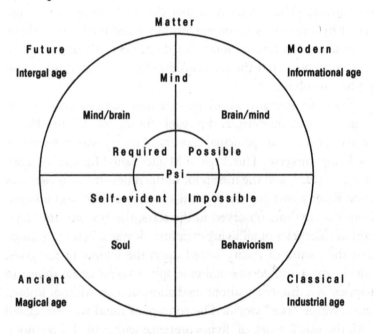

Figure 13–1. Evolution of physical and mental worldviews in relationship to psi.

The classical era evolved into the present, quantum era. While the origins of quantum theory can be traced to 1900, its impact on the world at large started to foment during the 1950s, took off in the 1980s, and, assuming present trends continue (which is always a shot in the dark), will reign until perhaps 2100. We might call the present era the "Age of Information," and the zeitgeist is a quantum *computer*. Basic physical concepts like space, time, energy, and matter are now imagined to be relative, complementary, and dependent in some ill-defined way on observation. Spooky actions at a distance are not merely possible, but required within our understanding of physical reality. The concept of mind is viewed as a dynamic, cybernetic interplay between a complex physical structure (the brain) and an emergent process (the mind), with brain imagined to be the primary driver of the process. Increasing numbers of scientists are beginning to ponder the role that quantum theory plays in the brain and in creating or sustaining consciousness. Psi is no longer flatly impossible. I believe this consensus will strengthen with time.

I anticipate that the present era may evolve into an "Integral Age," where the scientific worldview may revolve around holistic concepts. The zeitgeist will be the *noosphere,* Teilhard de Chardin's concept of Earth as a thinking organism. This era might begin around the middle of the twenty-first century and continue to blossom for the foreseeable future. Basic physical concepts will not just be viewed as complementary, but actively participatory. Mind will continue to be viewed as a dynamic interplay between brain and mind but also as more than an emergent process, perhaps as the primary driver of the process.

Just as psi was first taken for granted, then denied, and then allowed to exist again as each era came and went, theories of psi have paralleled the views of each era. In magical times, theories of psi were based on what we now regard as occult lore (Figure 13-2). Concepts like "astral" and "mental" bodies, elemental and divine spirits, and various forms of "lifeforce" were the prevailing ways that people imagined psi to be mediated. As supernatural magic evolved into natural magic, and alchemy and astrology evolved into chemistry and astronomy, concepts of psi began to evolve beyond stories based on invisible spirits. Some people today still use occult terms like "astral body" when referring to psi, but most scientifically minded researchers regard occult lore only as metaphors.

In the classical age, theories of psi followed advances in physics, thus ideas involving fields (as in "mind fields") and signal passing became popular. When the quantum age began, field theories (as in quasi-physical fields) and then quantum field–inspired theories were proposed. In the proposed integral age holistic and entanglement theories may become increasingly popular. Figure 13-2 is drawn in the form of a circle because one view of the evolutionary trend is that physics is returning to the holistic assumptions of the magical era. But there's an important difference. The post-modern view no longer lacks the explanatory power of the first era. In this sense the figure is actually not a circle but a spiral, winding up out of the page. Physics, and theories of psi, may be regarded as returning to an echo of the magical era, but in a drastically revised and far more precise form.

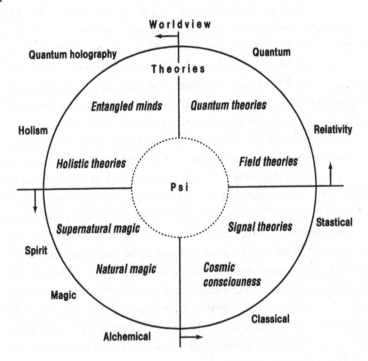

Figure 13–2. Diagram of theories of psi.

THEORIES OF PSI

In its broadest sense, a theory is a description of an observed effect. Theories can range from the explanatory precision of a mathematical equation to a metaphor or myth. The special power of scientific theories rests in their ability to make testable and therefore falsifiable predictions. Without having a way to test a theory, you can't tell if it's pointed in the right direction or not.

Theories of psi include the full range of possible descriptions, from mathematical to metaphorical, and from testable to nontestable.[2] Seven basic categories of psi theories can be formed. They fall into two types: Those offered to account for psi effects in general, and models attempting to account for specific effects in certain types of experiments. The following brief survey does not attempt to cover all published theories. Instead, I've identified a few theoretical *themes* that have been proposed, and I'll discuss representatives of those themes. The categories include:

- Skeptical theories
- Signal-transfer theories
- Goal-oriented theories
- Field theories
- Collective-mind theories
- Multidimensional space/time theories
- Quantum-mechanical theories

SKEPTICAL THEORIES

These theories attempt to explain reports of psi through a wide range of psychological frailties. This includes memory tricks, embellishment, wishful thinking, sensory illusions, implicit learning, underestimates of the frequency of coincidence, experimental design flaws, selective reporting (of anecdotes and experiments), weaknesses of eyewitness testimony, psychopathology, delusion, ignorance, and fraud.

There is no doubt that in the right context these factors can mimic psi effects. Indeed, if the only evidence for psi experiences were collections of anecdotes, it would be difficult to persuasively argue against such explanations. However, given that most of the experimental evidence discussed in previous chapters was designed to preclude such explanations, skeptical theories are insufficient as the sole explanation for psi.

SIGNAL-TRANSFER THEORIES

Signal-transfer theories propose that some sort of physical carrier wave, analogous to how electromagnetic (EM) waves carry radio signals, transports psi information. For many decades this was an appealing explanation for telepathy because we know that the brain generates EM fields, and we know that EM fields can carry information around the world at light speed. In 1899, the physicist Sir J. J. Thomson proposed in an address to the British Association for the Advancement of Science that EM fields might be the physical carrier of telepathy. The title of Upton Sinclair's book, *Mental Radio,* reflected the enthusiasm once held for signal-transfer models.

The most problematic issue for signal-transfer theories, assuming that the carrier is at least analogous to EM fields, is that for all known physical fields the strength of the field drops off quickly with increasing distance. If psi were mediated through any form of ordinary physical field, we'd expect psi accuracy to drop off rapidly with increasing distance. But psi experiments conducted under conditions of heavy EM shielding and at long distances do not show sharp declines in accuracy. This is not to say that the question of distance has been conclusively settled; as we've seen, there are inklings that distance (or possibly knowledge of distance) *may* play a role in some circumstances.

There is an exception to the rule about distance drop-offs. At extremely low frequencies (ELF, 0.3–1 kHz), EM field strength is sustained over long distances because it passes through barriers that would absorb or block higher frequencies. In the 1960s the Russian physicist I.M. Kogan proposed that telepathy is carried by ELF waves.[3] This was a novel solution to the problem encountered with higher frequency EM, but it too contains a problem. Like any method relying on EM, it cannot *easily* account for the apparent time independence of psi. So-called advanced wave solutions for EM could, in principle, allow for a form of retroactive signaling. But the time scales for such signaling would be limited to the speed of light, which means one could obtain a precognitive glimpse of only one nanosecond per foot as light travels. That is far too restrictive to account for precognition effects observed in life and lab, where future information can apparently be sensed from milliseconds to months or more in advance. It also doesn't easily account for retrocognition, the flip side of precognition, in which (hidden) information from the past can be perceived.

Signal-transfer theories have included proposals based on tachyons (hypothetical particles that travel faster than the speed of light), antimatter (particles that can be interpreted as traveling backward in time), neutrinos, and gravitons.[4] Unfortunately, all of these models suffer because the effects would be limited in space or time, and none of them account for clairvoyance. In addition, signaling theories don't provide a plausible explanation for telepathy, specifically in how signals sent from one brain might be "decoded" by another brain.

GOAL-ORIENTED THEORIES

These theories describe one of the main psychological characteristics of psi—its teleological or goal-oriented nature. They assume that psi "operates" so as to bring about desired goals. In this sense they're analogous to Aristotle's notion of different kinds of causes, in particular *efficient* and *final* cause. Efficient cause is the means by which we explain how everyday things work, like billiard balls colliding into other billiard balls, or intermeshing gears. Final cause assumes that end-goals are inextricably wound into the initial causes of events, or said another way, that *causes reverberate* between the starting and ending conditions of an event. These theories assume that outcomes in experiments do not depend heavily on the nature of the underlying physical system or the complexity of the experimental task.[5] These theories also assume that feedback is a critical element in producing psi effects because that's how the final goal is established.

Physicist Helmut Schmidt's version of a goal-directed theory was based on the quantum theory-inspired idea that observation influences probabilistic events.[6] Psychologist Rex Stanford's "Conformance Theory" was based on a model in which psi guides one's behavior and influences external events so as to achieve a goal.[7] Psychologist Michael Thalbourne's theory of psychopraxia, meaning self-achieving goals, was similar to that of Stanford.[8] And physicist Edwin May's theory, called "Decision Augmentation Theory," was based on the idea that if our decisions can be guided by precognition of possible futures, then we can optimize those decisions so as to achieve our goals.[9] All of these theories have made testable predictions; and overall the evidence suggests that they have merit, but it is not yet certain that feedback is a *necessary* ingredient for psi to "work."

FIELD THEORIES

Physical and quasiphysical field theories include psychiatrist Carl Jung's idea of the collective unconscious, biologist Rupert Sheldrake's morphogenetic fields, and neuroscientist Michael Persinger's geomagnetic field theory.[10] These models all postulate the existence of some form of nonlocal memory permeating time and space that we can resonate with. Jung and Sheldrake do not specify what these fields might consist of, and Persinger's model assumes that the Earth's geomagnetic field mediates the field. None of the theories suggest how *specific* information might be extracted from these fields other than through a resonance process. Of these theories, Sheldrake's idea has been experimentally tested, with some success.[11]

In a cognitive science approach to field theories, psychologist Christine Hardy has proposed a semantic fields theory.[12] She contends that because psi is intimately associated with the mind, and the mind's basis in physicality is still a hotly debated issue, there is (at present) no need to base a theory of psi exclusively on existing physical principles. Hardy proposes that the mind is "a lattice of semantic constellations . . . generated by the interplay of experience, genetic constraints and cultural context." She sees these constellations as self-organized, intertwined dynamic networks and psi events as the means by which these semantic networks are interconnected.

Other "mental field" models include William James's notion of a cosmic consciousness and Oxford scholar Frederic Myers's "metetherial world" or "subliminal self."[13] These theories assume that at some deep level individuals' minds are parts of a larger, unified mind, thus providing a natural explanation for experiences like telepathy and synchronicities. But these theories do not easily account for the broader range of psi experiences, including clairvoyance, precognition, and PK. And they do not lead to straightforward, testable predictions.

MULTIDIMENSIONAL THEORIES

Multidimensional theories are geometric approaches to solving the physics problem of how psi can transcend space and time. They first became popular in the latter part of the nineteenth century, when the concept of "the fourth dimension" captured the public's fancy. Many people were impressed by the idea that the time-space peculiarities of psi, which classical models could not accommodate, made a certain sense when a fourth spatial dimension was added. For example, the British psychologist Whately Carington published a book in 1920 entitled *A Theory of the Mechanism of Survival: The Fourth Dimension and Its Applications,* in which he argued that it might be possible for consciousness to survive in a fourth dimension.[14]

A recent and more sophisticated version of the multidimensional models is one by physicists Elizabeth Rauscher and Russell Targ. Their model assumes that the four-dimensional spacetime we're familiar with (three dimensions of space and one of time) is actually a complex eight-dimensional spacetime. Complex in this context refers to a branch of mathematics involving imaginary numbers, based on the square root of -1. The advantage of such a model is that it's consistent with all known physics, including quantum mechanics and relativity, and it provides zero distance in space or time between objects that appear to be separated. The model assumes that minds have the capability to navigate through the 8-space, and it's useful in showing how, in principle, one can account for the nonlocal properties required by psi without damaging known physics. However, it doesn't offer an explanation for how the required mental navigation might take place, or how PK might work.

QUANTUM THEORIES

Five theories of psi were inspired by quantum mechanics. While the word quantum is now used as an exotic adjective to augment the sales of everything from diets to fishing tackle, the connection proposed here is not trivial. As physicist Henry Stapp explains,

> Quantum approaches to consciousness are sometimes said to be motivated simply by the idea that quantum theory is a mystery and consciousness is a mystery, so perhaps the two are related. That opinion betrays a profound misunderstanding of the nature of quantum mechanics, which consists fun-

damentally of a pragmatic scientific solution to the problem
of the connection between mind and matter.[15]

Theory 1: Observational Theory

Observational Theory was proposed in the early 1970s. It was
motivated by similarities between the nonlocal properties of the
quantum wave function and the space-time independence of psi
phenomena, and also by the quantum measurement problem,
which offers the possibility that mind plays an important role in
physical reality. This theory is in accordance with the opinions
of Nobel laureates John Eccles and Eugene Wigner, and also
neuroscientist Wilder Penfield and mathematician John von
Neumann.[16] Wigner concluded from his own arguments about
symmetry in physics that the action of matter upon mind must
give rise to, as he put it, a "direct action of mind upon matter."[17]

Such radical ideas horrify traditionalists who prefer to think
of reality solely in classical terms. For example, in a January
2005 op-ed column in *Scientific American,* skeptic Michael Shermer expressed dismay over an interpretation of quantum theory
in which mind plays an active role in shaping reality, referring
to it as "quantum flapdoodle."[18] Perhaps Shermer missed a previous issue of *Scientific American* in 1979, in which physicist
Bernard d'Espagnat, writing about interpretations of quantum
theory, concluded that "The doctrine that the world is made up
of objects whose existence is independent of human consciousness turns out to be in conflict with quantum mechanics and
with facts established by experiment."[19]

Numerous researchers have made contributions to Observational Theory. The first formulation was by physicist Evan Harris Walker;[20] another early variation was proposed by physicist
Helmut Schmidt.[21] All of the variations of these theories assume
that the act of observing a quantum event probabilistically influences its outcome.[22] Observational Theory is particularly interesting because it led to a preposterous prediction: If random
data, like a series of random bits, were automatically recorded

on a computer hard disk without anyone observing them, then those recorded random bits would remain in an indefinite state until they were observed. *After* being observed, they would "collapse" into actual bits.

This prediction resulted in a series of experiments in which previously recorded, unobserved random bits were observed later according to instructions like "aim for 1s" or "aim for 0s." (It's important to emphasize that these instructions were generated *after* the bits were already recorded.) The results of the experiments were successful and consistent with the prediction that the act of observation *retroactively* influences quantum events.[23] Thus Observational Theory became one of the first theories of psi to predict and successfully confirm an outrageous time-reversed effect. Incidentally, the delayed-choice experiments discussed in the previous chapter provides *exactly* the same prediction as these "retro-PK" experiments. The only difference is that those experiments are considered mainstream in physics.

Theory 2: Model of Pragmatic Information

Physicist and psychologist Walter von Lucadou proposed that the basic structure of quantum theory might also be applicable to complex systems in general.[24] The motivation for this idea, as for all the quantum-inspired theories, is that quantum theory accurately accounts for observations ranging from subatomic to cosmological scales. Thus, it seems possible that the basic principles of quantum theory might apply in a more general way to basic relationships among information, space, and time.

Von Lucadou's model assumes that the *structure* and the *function* of a system—any system of any size or complexity—are complementary. That is, how a system is constructed, and how it behaves, are not merely related to one another. Instead, they are inextricably entangled. From this complementary relationship, von Lucadou proposes that an uncertainty relationship analogous to Heisenberg's uncertainty principle can be derived. This uncertainty relationship rests upon what von Lucadou calls "pragmatic information," or the *meaning* of the information. As with any uncertainty relationship, we can't measure both structure and function to arbitrary precision, because these properties are entangled. For example, if one attempts to precisely measure the *structure* or form of a bacterium by fixing it to a microscope slide, its *function* or behavior will be affected. And if one tries to determine the bacterium's function, then the means of making those measurements will likely change its structure. Likewise, Von Lucadou's model proposes that psi effects arise in nonlocal correlations that derive from the entangled structure-function relationship.

Theory 3: Weak-Quantum Theory

In an approach similar to von Lucadou's, psychologist Harald Walach proposed that "generalized entanglement" might be relevant to understanding psi.[25] This idea is an extension of an earlier proposal by Princeton researchers Robert Jahn and Brenda Dunne. They noted that physicist Neils Bohr and the other founders of quantum theory often wrote of complementarity as being a basic constituent of nature, including in the psychological domain.[26]

In the journal *Foundations of Physics* in 2002, physicists Harald Atmanspacher and Hartmann Römer, along with Walach, described an example of "Weak-Quantum Theory" in psychotherapy, specifically in the phenomenon of transference.[27] Transference refers to instances in which a client projects his or her problems onto the therapist, and countertransference to when the therapist projects his or her own issues back onto the client. Sometimes aspects of the client's life that were not consciously known (to the client) arise in the thoughts of the therapist, and vice versa. Atmanspacher and his colleagues proposed that weak quantum theory predicts such "entangled mental states" due to the complementarity or entanglement of shared conscious and unconscious states. As with other complementary conditions, the uncertainty between these mutually exclusive conditions creates nonlocal connections, in this case between the "entangled" client and therapist.

Weak-Quantum Theory also notes other complementarities where nonlocal connections may potentially occur, including mass and energy, space and time, waves and particles, fields and quanta, real and imaginary numbers, zero and infinity, analysis and synthesis, organic and inorganic, and in general, parts and wholes.

Theory 4: Bohm's Implicate/Explicate Order

Einstein's protégé, American physicist David Bohm, felt that quantum theory suggested the existence of a deeper reality than the one presented by our senses. He dubbed the *implicate order* an undivided holistic realm that is beyond concepts like space-time, matter, or energy. In the implicate order everything is fully enfolded or entangled with everything else. By contrast, the *explicate order* world of ordinary observations and common sense emerge, or unfold, out of the implicate order.

Bohm used a hologram as a metaphor to illustrate how information about a whole system can be enfolded into an implicit structure, any part of which reflects the whole. From this perspective, when it comes to human experience Bohm wrote:

> It will be ultimately misleading and indeed wrong to suppose, for example, that each human being is an independent actuality who interacts with other human beings and with nature. Rather, all these are projections of a single totality . . . .[28]
> In the implicate order we have to say that mind enfolds matter in general and therefore the body in particular. Similarly, the body enfolds not only the mind but also in some sense the entire material universe . . . . We have evidently to include matter beyond the body if we are to give an adequate account of what actually happens and this must eventually in-

clude other people, going on to society and to mankind as a whole. . . . [29]

Stanford neuroscientist Karl Pribram independently proposed a concept similar to Bohm's ideas about a quantum holographic reality but applied to processes in the human brain. In examining the structure and functioning of the brain, Pribram was struck by similarities in how the brain and optical holograms store information. While holograms are not dynamic processors like the brain, the basic idea, according to Pribram, bears a certain resemblance:

> In the brain, when we look at the electrical impulses traveling through the neurons, and the patterns as these billions of neurons interact, you would say that that is analogous . . . to the processes that are going on at the deeper quantum level. . . . If indeed we're right that these quantum-like phenomena . . . apply all the way through to our psychological processes, to what's going on in the nervous system—then we have an explanation perhaps, certainly we have a parallel, to the kind of experiences that people have called spiritual experiences. Because the descriptions you get with spiritual experiences seem to parallel the descriptions of quantum physics.[30]

These twin ideas—Bohm's holographic universe and Pribram's holographic brain—were popularized by author Michael Talbot in his book, *The Holographic Universe*. In it, Talbot proposed that a combination of Bohm's and Pribram's ideas could account for a vast range of paranormal and psychic experiences. Similar proposals have been discussed in books edited by psychologist Ken Wilber.[31] The holographic paradigm is now being used by cosmologists to mathematically model the physical structure of the universe,[32] and the concept of reality as a *quantum hologram,* a self-referencing system based on the interference properties of quantum waves, is beginning to attract attention.[33]

As noted in a news clip on the web site of the American Institutes of Physics:

> Second sight and remote viewing are terms used to explain charlatans' supposed psychic ability to see hidden objects in terms of pseudoscientific gibberish. Quantum holography, on the other hand, is a method firmly grounded in modern physics that permits the imaging of hidden objects with entangled photons.[34]

Reading between the lines, one gets the sense that the American Institutes of Physics is somewhat biased against the concept of psi. I suspect this may have more to do with concerns over public image rather than with any real substance.

Theory 5: Stapp-von Neumann

In 1932, the eminent Hungarian mathematician John von Neumann placed the foundations of quantum theory on firm mathematical ground; since then his formulation has been considered the orthodox "core" of quantum theory. Von Neumann's interpretation, like the Copenhagen interpretation, assumes that quantum theory tells us about the observer's *knowledge* of reality rather than "Reality" itself, and that the observing instrument and what is observed is part of a whole system. Physicist Henry Stapp, of the Lawrence Berkeley National Laboratory, has recently refined von Neumann's interpretation.[35] The Stapp-von Neumann approach assumes that because a key component in the quantum measurement process includes an observer and his or her knowledge, this means the mind is inextricably wound into quantum reality. While this was not proposed as a theory of psi, it seems to lead naturally to that conclusion. Let's see how.

Stapp asserts that a key advantage of the von Neumann approach is that it overcomes a limitation in how consciousness is understood within classical physics. Based on the classical assumptions of local realism and mechanism, the brain—like any other physical object—is a clockwork object. Since clockworks are not conscious, then what we call "I" can only be an emergent property of a complicated piece of machinery. And thus our sense of conscious awareness, or the feeling one has when smelling a rose, are illusions—though illusions to *whom* is not quite clear. From a classical physics point of view, the "you" that is currently reading this sentence is an illusion. This seems to be a rather important limitation, as most people reading these sentences probably believe that they (their conscious minds) do exist.

The Stapp-von Neumann approach solves this problem by putting the mind back into the quantum-measurement process. It proceeds through two events, dubbed Process I and Process II. In simplified form, Process I involves the mind asking a question of Nature, and Process II is her response.[36] Process I probes Nature from "outside" the usual constraints of space and time (and is thus a nonlocal process), whereas Process II is what we observe within Nature, and thus limited by ordinary space-time. As Stapp explains,

> [This reveals] the enormous difference between classical physics and quantum physics. In classical physics the elemental ingredients are tiny invisible bits of matter that are idealized miniaturized versions of the planets that we see in the heavens, and that move in ways unaffected by our scrutiny, whereas in quantum physics the elemental ingredients are intentional actions by agents [i.e., minds], the feedback arising from these actions, and the effects of our actions on the physical states that embody or carry this information.[37]

What does all this have to do with psi? It suggests that the mind/brain might be a self-observing quantum object, and as such, it resides within an entangled, nonlocal medium that just happens to be entirely compatible with the known characteristics of psi. The brain is enormous in comparison to individual quantum objects like atoms. So how does the mindlike Process I interact with the evolving state of the brain? Some have proposed that structures within the neurons, called microtubules, may be able to sustain quantum effects within the brain.[38] Stapp, like physicist Evan Harris Walker, offers a somewhat different answer—the state of the brain is highly sensitive to events occurring at the atomic level, in particular at the boundaries between neurons—the synapses.

Neurons communicate with each other through the release of neurotransmitter molecules. When an electrical signal reaches the end of a neuron, it causes channels in the neuron to open through which calcium ions can enter. If a sufficient number of ions are accumulated, the neuron releases neurotransmitters, which in turn increase (or sometimes decrease) the tendency of surrounding neurons to "fire" their own electrical signals. Multiply that process by a few billion neurons and trillions of synapses, and that's the basic communication infrastructure of the brain.

The quantum element enters at the ion channels, because at some points these channels are less than a billionth of a meter (a nanometer) in diameter, and at that size quantum effects become noticeable. Stapp proposes that the quantum uncertainty of the ion's location causes it to "spread out" and become a cloud of potentials instead of a classical particle at a specific location. This means that both *where* (and also *whether*) the ion lands on a given neurotransmitter triggering site is indeterminate. And there are trillions of brain locations where this is occurring, continuously.

This paints a picture of a dynamically changing ensemble of quantum ion probability clouds, most of which are "observed" by the brain itself and consequently collapse into particles through the process of quantum decoherence (interactions within the environment of the brain). Thus most brain processes churn away comfortably without a conscious mind directing the show, as many neuroscientists believe. So why is the mind needed? Stapp suggests:

> The brain is warm and wet, and is continually interacting strongly with its environment. It might be thought that the strong quantum decoherence effects associated with these conditions would wash out all quantum effects. [But] because of the uncertainties introduced at the ionic, atomic, molecular, and electronic levels, the brain state will develop not into one single classically describable macroscopic state, as it does in classical physics, but into a *continuous distribution of parallel virtual states* of this kind.

So the conscious mind is needed to direct this dynamic distribution of states into a single state of focused awareness. Otherwise the brain would operate more like a diffusely daydreaming cauliflower than a thinking, conscious organ. To provide this direction, the mind takes advantage of the fact that the dynamic state of the brain often comes to a tipping point where it must decide between two or more different responses. This provides an exquisitely sensitive pivot to query the ion probability clouds by Process I [the mind] so as to cause an ion cloud to collapse on one neuron's receptor site rather than another.

How does the mind/brain cause one particular line of thought, or decision, to be sustained over another? Stapp offers an intriguing speculation based on the Quantum Zeno Effect.[39] This refers to a prediction (since confirmed by experiments) that the act of rapidly observing a quantum system forces that system to remain in its wavelike, indeterminate state, rather than to collapse into a particular, determined state. As Stapp says,

> Taken to the extreme, observing continuously whether an atom is in a certain quantum state keeps it in that state for-

ever. For this reason, the Quantum Zeno Effect is also known as the watched pot effect. The mere act of rapidly asking questions of a quantum system freezes it in a particular state, preventing it from evolving as it would if we weren't peeking. Simply observing a quantum system suppresses certain of its transitions to other states.

This means that if the dynamic state of the brain is repeatedly self-observed it will tend to sustain certain brain states more than others.[40] And this is what Stapp views as the mind "directing the show" with attention and intention. In this sense, what we call "attention" is explained as a consequence of the brain applying the Quantum Zeno Effect to itself. Likewise, what we call "intention" is the act of directing that attention toward some goal.

Thus, the Stapp-von Neumann approach to quantum mind allows for the mind to choose among different brain states. This does not imply that brain and mind are necessarily different "substances." The mind can be thought of as that portion of the brain that observes and directs itself. Regardless of whether we conceive of von Neumann's Process I as a dualistic interaction between a mind *and* a brain, or as a unitary mind/brain process, the process itself is defined as *nonlocal.* This opens the possibility that one person's mind/brain can cause the probabilistic brain states of another person or another object (or other human organs, like the gut), to preferentially collapse into selected states. And that opens the door for psi.

ENTANGLED MINDS

Hence this life of yours which you are living is not merely a piece of the entire existence, but is, in a certain sense, the whole; only this whole is not so constituted that it can be surveyed in one single glance.

—Erwin Schrödinger[41]

For entangled minds to accurately describe and predict psi performance, we'll need a model that combines features of physics, neuroscience, and psychology. For physics, we must reside in a medium that supports connections transcending the ordinary boundaries of space and time. For neuroscience, minds (by which I mean mind/brains) must be sensitive to and play an active role in that medium. And for psychology, the processes of attention and intention should play key roles in how mind "navigates" within this medium.

The first issue is whether the fabric of reality allows for nonlocal connections. As we've seen, this question has been answered in the affirmative for 80 years theoretically and for 20 years experimentally. Quantum theory successfully describes physical behavior from the atomic to cosmological domains, with no experimental violations observed to date. It would be astonishingly unlikely to find that one small domain, the one that our bodies and minds happen to inhabit, are somehow *not* best described as quantum objects. As historian of science Robert Nadeau and physicist Menas Kafatos, both from George Mason University, describe in their book *The Nonlocal Universe*:

> All particles in the history of the cosmos have interacted with other particles in the manner revealed by the Aspect experiments. Virtually everything in our immediate physical environment is made up of quanta that have been interacting with other quanta in this manner from the big bang to the present . . . .
>
> Also consider . . . that quantum entanglement grows exponentially with the number of particles involved in the original quantum state and that there is no theoretical limit on the number of these entangled particles. If this is the case, the universe on a very basic level could be a vast web of particles, which remain in contact with one another over any distance in "no time" in the absence of the transfer of energy or

information. This suggests, however strange or bizarre it might seem, *that all of physical reality is a single quantum system* that responds together to further interactions.[42]

It's tempting to assume that quantum reality plays no role when it comes to understanding phenomena like human experience. But the fact is that we don't know how "big" an influence has to be to cascade our brain states into one set of subjective experiences versus another. If Stapp and others are correct about the quantum mind/body connection, then human experience is indeed a part of quantum reality. As Nadeau and Kafatos put it,

> We can no longer rationalize [quantum] strangeness away by presuming that it applies only to the quantum world. Bohr was correct in his assumption that we live in a quantum mechanical universe and that classical physics represents a higher-level approximation of the dynamics of this universe. If this is so, then the epistemological situation in the quantum realm should be extended to apply to all of physics.[43]

As an aside, it's interesting to note that Nadeau and Kafatos mention early in their book that readers accidentally encountering their book in the "new age" section of a bookstore would likely be disappointed. That's because the book is about physics and not new age ideas. But the fact that Nadeau and Kafatos felt it important to mention this at all illustrates the rising tension between the leading edge of interpretations in physics and the tail end of metaphysics. Physicists interested in quantum ontology are painfully aware that some interpretations of quantum reality are uncomfortably close to mystical concepts. In the eyes of mainstream science, to express sympathy for mysticism destroys one's credibility as a scientist. Thus the taboo persists.

INSIDE THE ENTANGLED MIND

To see the world in a grain of sand
And a heaven in a wild flower,
Hold infinity in the palm of your hand
And eternity in an hour.

—William Blake

Blake's poem hints at how an entangled mind might perceive the world. Another poetic description can be found in Frank Herbert's description of the visionary protagonist, Paul Muad'Dib, in the science fiction novel, *Dune*. When Muad'Dib takes the mind-opening drug called *spice* his perceptions transcend time and space. This passage describes an episode from one of his visionary experiences:

> The prescience, he realized, was an illumination that incorporated the limits of what it revealed—at once a source of accuracy and meaningful error. A kind of Heisenberg indeterminacy intervened: the expenditure of energy that revealed what he saw, changed what he saw.
> And what he saw was a time nexus . . . a boiling of possibilities focused here, wherein the most minute action—the wink of an eye, a careless word, a misplaced grain of sand—moved a gigantic lever across the known universe. . . . The vision made him want to freeze into immobility, but this, too, was action with its consequences.

My guess is that Herbert's and Blake's descriptions are pointing in the right direction for an understanding of psi. At a level of reality deeper than the ordinary senses can grasp, our brains and minds are in intimate communion with the universe. It's as though we lived in a gigantic bowl of clear jello. Every wiggle—every movement, event, and thought—within that medium is felt throughout the entire bowl. Except that this particular form of jello is a rather peculiar medium, in that it's not localized in the usual way, nor is it squishy like ordinary Jell-O. It extends beyond the bounds of ordinary spacetime, and it's not even a substance in the usual sense of that word.

Because of this "nonlocal Jell-O" in which we are embedded, we can get glimpses of information about other people's minds, distant objects, or the future or past. We get this not through the ordinary senses and not because signals from those other minds and objects travel to our brain. But because at some level our mind/brain is *already coexistent* with other people's minds, distant objects, and everything else. To navigate through this space, we use attention and intention. From this perspective, psychic experiences are reframed not as mysterious "powers of the mind" but as momentary glimpses of the entangled fabric of reality.

Particles that are quantum entangled do not imply that signals pass between them. Entanglement means that separated systems are *correlated*. Psi, on the other hand, seems to involve information transfer, like signal passing. At first glance, that seems to eliminate quantum correlations as an explanation of psi. However, the pseudo-telepathy paradigm discussed in the previous chapter shows that joint tasks that would require classical signals can take place *without* any information transfer. This suggests an alternative understanding of psi. Maybe it doesn't involve information transfer at all. Maybe it's purely relational and manifests only as correlations.

To explain this in more detail, let's assume that our bodies, minds, and brains are entangled in a holistic universe. It is not necessary to assume that the mind is fundamentally different from the brain, or the even more radical notion that reality is created by consciousness. It is only necessary to imagine that the mind/brain behaves as a quantum object. Imagine that our mind/brain is sensitive to the dynamic state of the entire universe. There are an astounding number of events we can potentially react to, but the vast majority of them can be regarded as background noise. Other than where your body is, you might be interested in perhaps ten other locations or events within the universe at any given moment, all of them relatively close to you in spacetime.

Some portion of your unconscious mind pays attention to those selected locations at all times. Like suddenly hearing your name mentioned at a noisy cocktail party, you become consciously aware of items of interest through your unconscious scanning ability. Most of your *conscious* awareness is heavily driven by sensory inputs. That sensory-bound brain state is also entangled and influenced by the rest of the universe, but its local effects are so much stronger and immediate than our "background" awareness that only on rare occasions are we aware of its entangled nature. A few gifted individuals are able to direct their conscious awareness at will to surf through the entangled unconscious, but even they have trouble maintaining that state for long. For as the fictional character Paul Muad'Dib described, the act of seeing disturbs that which is seen. For the rest of us, we have to rely on our unconscious mind(s) to pay attention to those fleeting events of interest.

On occasion, if a distant loved one is in danger, the part of your unconscious that has been attending to the environment alerts your conscious self. You might experience this alert as a gut feeling, as an odd sense of something meaningful afoot, or your imagination might be activated and you might perceive a fleeting vision of your loved one. On extraordinary occasions, you might obtain a veridical sense of what is happening somewhere else. That vision is a construction from your memory and imagination, similar to a waking dream except that the stimulus for this image is occurring somewhere, or some-when, else.

If you later learned that indeed your loved one was in danger, or wished to communicate with you, then you'd call this a case of spooky telepathy. It would appear to be a form of information transfer, but in fact it would be a pure correlation. That is, within a holistic medium we are *always connected*. No information transfer need take place because there are no separate parts. Navigation through this reality occurs through our attention, and nonsensory perception takes place through our activated memory and imagination.

QUESTIONS ABOUT ENTANGLED MINDS

How it is possible to gain information without the use of the ordinary senses, and not bound by the usual constraints of space and time? The brain, like all other objects, is part of the entangled fabric of reality. As such, brain functioning is not just ruled by classical physics and biochemistry, but also participates in events distributed throughout space and time. Events may be thought of as ripples reverberating throughout an immense pool, and the brain as an object bobbing along the surface like a cork. Nonsensory perceptions are occasionally evoked in the brain because, as an exquisitely sensitive pattern recognizer, it responds to ripples resembling similar undulations associated with previous events. So similar memories arise. If the unconscious mind deems those memories to be sufficiently interesting, then information will arise to awareness in the form of imagination or fleeting thoughts.

One implication of the bobbing brain idea is that we wouldn't be able to perceive something via psi that we weren't already familiar with. If we asked a psychic to clairvoyantly describe what was happening on Mars 12 million years ago, in principle she should have access to that information. But even if we asked her to describe this target blindly, so she wasn't biased by prior expectations, she would still be limited to perceiving Earthlike, familiar settings, as that's what available in her memory. Thus, if she described blue humanoids ambling through a suburban Martian shopping mall, it would be a mistake to assume that that perception was veridical.

Truly alien events and places, which means the vast majority of the universe and some restaurants in Los Angeles, would be so foreign to the mind that such "perceptions" would likely never rise to conscious awareness at all. The flip side of this implication is that objects we are most familiar with, like our loved ones, local environment, and people and places that are most meaningful to us, would be most likely to be perceived with fidelity and reach our awareness. This may be why the vast proportion of spontaneous psi experiences are of people and places that are especially meaningful to us.

Why is psi so elusive in laboratory tests? French philosopher and Nobel laureate Henri Bergson gave the presidential address to the Society for Psychical Research in London in May 1913. In that address, he proposed that one function of the brain was to enable conscious awareness to be held "fixed on the world in which we live." Bergson conceived of the brain as a filter, protecting consciousness from being overwhelmed by excessive stimulation, so we may focus on physical survival.[44] He added,

> If telepathy is a real fact, it is very possible that it is operating at every moment and everywhere, but with too little intensity to be noticed, or else it is operating in the presence of obstacles which neutralise the effect at the same moment that it manifests itself. We produce electricity at every moment, the atmosphere is continually electrified, we move among magnetic currents, yet millions of human beings lived for thousands of years without having suspected the existence of electricity. It may be the same with telepathy.[45]

From this perspective, psi is weak because the brain/mind has evolved to filter out awareness of most of the external world. If this were not so, then even most sensory information would be overwhelmingly distracting. This filtering process also includes awareness of events elsewhere in space and time, as those perceptions would be vastly more overwhelming than ordinary sensory inputs. In addition, the entangled universe is not simply an enormously complex system, it is also exquisitely reactive to both actions and observations. That recursive relationship guarantees that psi will be elusive. It's like looking at your eye in a mirror to remove an eyelash. The moment you shift your eye to get a better look at the eyelash, the whole image moves.

Maybe the universe was entangled for the first few nanoseconds after the Big Bang, but how could it have remained entangled for billions of years afterwards? Einstein's Special Theory of Relativity proposed that matter and energy are different aspects of the same substance, and the atomic bomb confirmed that proposal. Thus entanglement is a property of both matter (as in atoms) *and* energy (as in photons). This means that the bioelectromagnetic fields around our bodies are entangled with electromagnetic fields in the local environment *and with* photons arriving from distant stars. The brain's electromagnetic fields are entangled with the rest of the universe not because of direct contact, in the sense of billiard balls colliding, but because its fields interpenetrate with the energetic fields of everything else. This is also how the universe remains entangled.

Why is psi often goal-oriented, and why does meaning sometimes amplify psi effects in life and lab? We are motivated by psychological intentions and organic needs, so psi, which is mediated through conscious and unconscious psychological drives, strongly reflects those needs. Also, a good portion of the individual brain/mind is engaged in "meaning creation," that is, making sense of its perceptions. So entangled minds will also be intimately involved in meaning creation and modulated by psychological beliefs and our need for meaning.

Why are psi missing (significantly avoiding the correct target), displacement (accurately describing nearby objects instead of the selected target), and decline effects (results that decrease with repeated efforts) sometimes observed in psi experiments? I suspect that these effects are due to the psychological filters through which psi manifests. Note that "psychological" does not necessarily mean limited to one individual. The entangled-minds concept assumes that individual beliefs and desires are not strongly localized, thus if a strong psi effect is observed and widely reported, that knowledge may produce a collective reaction among those groups who wish for the effect to go away (a societal "immune response"). This, in turn, will make it progressively more difficult to sustain the effect.

Psi missing occurs because the conscious mind wishes to avoid certain experiences. Psi missing has been observed most commonly in studies involving comparisons in psi performance between sheep (believers) and goats (nonbelievers). The goats do not want to see evidence for psi, so they tend to systematically hit below chance expectation, thus supporting their desire.

Displacement occurs because minds are entangled not only with the desired target image (in say, a telepathy experiment) but with all possible targets. If a target *pool* is especially meaningful or interesting in an experiment, and if feedback further entangles the mind to all of the targets (as is common in the ganzfeld telepathy experiment), then the mind can become confused as to which target is the "important" one. This may also be related to the psi-missing effect, in that displacement effects often occur after a string of exceptionally good hits. The experience of strong psi effects can evoke a repression response that deflects the conscious mind from naming the correct target.

Decline effects occur in many types of experiments that employ repeated trials. The principle culprit is probably boredom. Novelty is important in maintaining the high attention required to sift entangled wheat from chaff. When boredom sets in, it's unlikely that attention can be maintained, and thus performance declines.

Why does psi spontaneously occur more often in altered states of consciousness, like dreams and meditation? The ordinary waking state is largely driven by sensory awareness, so anything that disrupts that awareness will probably improve psi perception. This may be why people with unstable temporal lobes report more psi, and why traditional shamanic methods of creating nonordinary states, like meditation, drumming, chanting, and psychoactive drugs are associated with reports of improved psi perception. People with natural psi talent do not appear to require special states of consciousness, although their ability to quickly shift between mental states may be what defines their talent.

How would entangled minds explain the evidence for collective forms of consciousness, as in field consciousness effects? Minds are entangled with the universe, so in principle minds can nonlocally influence anything, including a collection of other minds or physical systems. Individual neurons in the brain combine into networks of neurons, giving rise to complex brain circuits and conscious awareness (or correlates of awareness). By analogy, individual minds may combine into networks of entangled minds, giving rise to more complex "mind circuits," forms of awareness, and collective psi effects beyond our conception.

How do mind-matter interaction (psychokinetic) effects work? In an entangled medium minds and intention are not just located *here*–minds are everywhere and every-when. Anything that resides, even momentarily, in a quantum indeterminate state may be susceptible to influence from nonlocal minds. This predicts that the more inherent indeterminacy there is within an object, the more likely it can be influenced via thought (PK). Thus, it should be more difficult to mentally affect a rock than a bacterium.

Large-scale effects like levitation or teleportation may be possible in principle, but the laboratory evidence for such claims is quite poor. If someday such effects are credibly demonstrated, one way they might be possible is if mind could influence energetic equilibrium states, even to miniscule degrees. For example, if mind could reduce the atmospheric pressure under a soda can just slightly, then using the same principle of unbalanced air pressure that makes an airplane wing fly, that unbalanced pressure would shoot the can up a few dozen feet before the equilibrium was reestablished. Likewise, if mind could momentarily alter the energetic equilibrium of the quantum zero-point field under a soda can, which is very roughly speaking the energetic equivalent of air pressure, then before the equilibrium state rebalanced it might launch the can into orbit.

Quantum correlations do not involve signal transfer, but psi appears to require signaling. So is quantum entanglement really a viable model for psi? Biological systems are clever in figuring out ways of using inanimate matter in ways that would not be predicted based on the properties of those materials alone. So living systems may be able to figure out how to use quantum correlations to communicate. Short sequences that appear to be random on the atomic scale would have enormous meaning to living systems. At the atomic scale, the dot of ink above the first "i" in the sentence, "You have won a million dollars," is distributed more or less at random from the perspective of the atoms on the rest of the page. But at our level, when we see this dot in context, it is not random and, indeed, it acts as a catalyst that creates a massive energetic effect. That energetic release could not be predicted from the viewpoint of the atoms in the "i."

Physicist Brian Greene says that while he likes the sentiments of an entangled universe, "such gushy talk is loose and overstated."[46] *So isn't a theory of psi based upon entanglement merely a panacea that predicts little more than "everything is everything"?* Entanglement left over from the Big Bang is analogous to a low-level, background radiation. We are indeed entangled with everything, so in principle we can (mentally) interact with everything and anything. But since entanglement increases in proportion to the number of interactions, it's conceivable that we're more likely to perceive information that's local to us in spacetime than events that happened a million years ago or a million light-years away.

This could be tested by performing telepathy tests with family members vs. strangers, and predicting better results with the higher entangled objects, namely family members (there is some evidence supporting this idea). But because everything on Earth is already entangled to a high degree due to the extensive exchange of atoms and electromagnetic fields on the globe, we might have to conduct psi tests between objects here and objects on other planets to observe strong differences in performance.

We could also predict that to improve psi performance in a telepathy experiment we should enhance the degree of entanglement, perhaps by using identical twins who spend a great deal of time with each other; we should limit who knows about the ongoing experiment and its results to help constrain how much nonlocal mental "noise" is interacting with the experiment; we should test twins who have excellent memories, are experienced at generating imagery, are both open to psi, and have the ability to maintain high levels of concentration for extended periods of time; we should use a task that is novel and highly motivating; and the twins should be selected based on their having prior telepathic experiences.

The closest we have to actual experiments testing this model are studies involving talented creative arts students in the ganzfeld telepathy experiments. For that subset of participants, the direct hit rates are 50% to 75% where chance expectation is 25%. So there's reason to believe that much higher psi performance is possible than is usually observed in lab experiments.

If psi is real, why hasn't it been further developed by evolution given the apparent advantages it would provide? It might well be the case that evolution has taken advantage of psi, but we haven't noticed it yet. For example, physicist Johann Summhammer showed in a March 2005 paper entitled "Quantum Cooperation of Insects" that if insects shared entangled states they could accomplish tasks more efficiently than if they had to rely on classical forms of communication.[47] In his analysis, Summhammer used an example of two ants pushing a pebble that was too heavy for one ant, and a second example of two distant butterflies that wanted to find each other. He showed that two quantum-entangled ants could push the pebble up to twice as far as two classical ants, and two entangled butterflies could find each other up to 48% faster than two classical butterflies. Based on this analysis, he proposed that if biological systems *were* entangled, then because of the advantages it provides, evolution may well have found a way to use it.

From another perspective, it's feasible that a mutant human might come along every now and then who was exceptionally sensitive to the entangled universe. The question is whether this mutation would be sufficiently compatible with normal psychological functioning for it to survive. For example, genius level intelligence would seem to offer an important evolutionary advantage for the individual and for society. So why aren't we all geniuses by now? One answer is that some advantages are self-extinguishing. Genius can come uncomfortably close to madness, and madness does not offer a survival advantage. Likewise, exceptionally strong natural psi ability might *seem* to provide a survival advantage, but it might also carry a tendency towards psychological dissociation, toward hypersensitive levels of empathic identification, and so on. Some forms of schizophrenia might be due to brains that are overloaded by psi information.[48]

In a society that seeks out and cultivates people with natural psi talent, and cares for their special sensitivities, it's conceivable that groups with refined psi abilities could prosper. Such groups might prove to be extremely useful to society. Unfortunately, it's also likely that the existence of such groups would introduce intense fear and resentment in those who were less gifted, and it isn't clear that such a group could be controlled for very long by "outsiders." Thus if such groups *were* formed, they'd have to be established under conditions of extreme secrecy. This is a favorite science fiction theme that, like much in science fiction, might have a grain of truth in it.

CHAPTER 14

NEXT

Prediction is very difficult, especially about the future.

—Niels Bohr

THE STATE OF THE ART

After a century of increasingly sophisticated investigations and more than a thousand controlled studies with combined odds against chance of 10^{104} to 1 (Table 14–1), there is now strong evidence that some psi phenomena exist.[1] While this is an impressive statistic, all it means is that the outcomes of these experiments are definitely not due to coincidence. We've considered other common explanations like selective reporting and variations in experimental quality, and while those factors do moderate the overall results, there can be little doubt that overall something interesting is going on. It seems increasingly likely that as physics continues to refine our understanding of the fabric of reality, a theoretical outlook for a rational explanation for psi will eventually be established.

EXPERIMENTAL CLASS	STUDIES	TRIALS	ODDS AGAINST CHANCE
Dream psi	47	1,270 sessions	2.2×10^{10} to 1
Ganzfeld psi	88	3,145 sessions	3.0×10^{19} to 1
Conscious detection of being stared at	65	34,097 sessions	8.5×10^{46} to 1
Unconscious detection of distant intention	40	1,055 sessions	1,000 to 1
Unconscious detection of being stared at	15	379 sessions	100 to 1
Dice PK	169	2.6 million dice tossed	2.6×10^{76} to 1
RNG PK	595	1.1 billion random events	3,052 to 1
Combined	1,019		1.3×10^{104} to 1

Table 14–1. A meta-meta-analysis of the classes of experimental evidence considered in this book. The number of studies listed, and the odds against chance, are adjusted for potential selective reporting biases using the trim and fill algorithm. The combined results indicate that these experimental results are unlikely to be due to coincidence or dumb luck. Something else is going on. Genuine psi offers an increasingly plausible interpretation.

SO WHAT?

What difference does it make if psi is real or not? I believe the principal effect that a scientific acceptance of psi would provoke in the short run is a change in worldview. Real psi carries profoundly important implications for our understanding of who and what we think we are. It identifies an entirely new realm of knowledge. It would have the same type of impact as discovering life on other planets, or finding evidence of advanced civilizations living on Earth 20,000 years ago, or a UFO landing on the White House lawn.

It would also force us to reevaluate ancient lore about the nature of consciousness itself. Over thousands of years, Eastern meditative practices have been used to trace what happens to the mind during sleep and during the transition to death. The Tibetan tradition of dream yoga, and the extensive literature on the *bardos,* the transitional states between living and dead (and beyond), suggest that the Western scientific understanding of life and mind may have been examining only a tiny portion of our capabilities.[2] As a famous Sufi parable teaches, it's as though we've lost our house key somewhere on the road, but we've only been looking for it near the streetlamp because that's where the light is. Perhaps we've been seduced by our tools to only look in certain spaces, and in the process we've overlooked something very interesting.

Virtually all meditative traditions take for granted that what we call psi is simply the initial stages of awareness of deeper levels of reality. If psi can be confirmed using Western scientific methods, then what shall we make of the rest of Eastern lore? Does some aspect of mind survive bodily death? Are there other forms of existence? Other intelligences? In the West, these sorts of questions have been relegated exclusively to the province of religion and superstition. But perhaps they can be probed with increasingly refined scientific methods, without invoking fear and ignorance of the unknown.

SKEPTICISM

In spite of the evidence, many remain skeptical. There's nothing wrong with this attitude; doubt is healthy. But extreme skepticism is another matter. This is not the place to examine the psychology of hyper-skepticism, but it's difficult to overlook the fact that fanatically skeptical groups seem to be motivated more by anger and cynicism than by a dispassionate search for the truth.

Reasonable doubt is sustained by three related factors: First, it's quite true that no one has developed a foolproof recipe that can guarantee a successful psi experiment conducted by anyone at any time. But it's also true that after spending say, a hundred billion dollars on cancer research, no one can guarantee a successful recovery from most types of cancer, or even a successful diagnosis. The reality is that some problems are exceedingly difficult, and psi is one of those problems. If we imagined that all funds raised for cancer research were spent in a single day, then the comparative funding for psi research—all of it, worldwide, throughout history—is conservatively equivalent to what cancer research consumes in a mere 43 seconds.[3] From this perspective, it's amazing we've learned anything at all, and it suggests that psi may be more pervasive than we've imagined. It's just difficult to disentangle ourselves from our environment to see psi clearly; we are like fish assigned to study the nature of water.

Second, most scientists aren't aware of the relevant experimental literature, nor have many paid attention to the ontological changes that have been reshaping the foundations of science. While articles on psi-related topics appear in mainstream journals from time to time, they are vastly outweighed by more conventional work, so advancements in psi research are easy to overlook. Also, scientific disciplines are so specialized today that no one can be expected to be familiar with more than one tiny slice of available knowledge. So maintaining doubt about remarkable claims in other disciplines is perfectly reasonable.

But the third, and in my estimation principal reason for persistent skepticism is that scientific truths do not arise solely through the accumulation and evaluation of new evidence. In particular, consensus opinion advances through authoritative persuasion. This is not how it's *supposed* to work in an ideal world, but the fact is that editorials by scientists in prominent magazines and newspapers regularly sway both public and scientific opinion. Use of rhetorical tactics like ridicule are especially powerful persuaders in science, as few researchers are willing to risk their credibility and admit interest in "what everyone knows" is merely superstitious nonsense.

That such persuasions influence consensus "truth" is anathema to the spirit of scientific exploration, but there's no doubt that it occurs. Since the turn of the twenty-first century, such devices have become flamboyantly obvious. There have always been backroom cartels between politics, business, and science, but today prominent medical scientists are openly paid "consulting" fees by pharmaceutical companies for promoting their products.[4]

In earlier times, when business and science were less deeply enmeshed, many scientists felt that a mass of new evidence could, at least in principle, sway consensus opinion toward a controversial idea. For example, in London in 1882, Henry Sidgwick gave the first presidential address to the Society for Psychical Research. Sidgwick was a Cambridge professor of ethics and a prominent moral philosopher. At that meeting he said:

> It is a scandal that the dispute as to the reality of these [psi] phenomena should still be going on, that so many competent witnesses should have declared their belief in them, that so many others should be profoundly interested in having the question determined, and yet that the educated world, as a body, should still be simply in the attitude of incredulity.
>
> Scientific incredulity has been so long in growing, and has so many and so strong roots, that we shall only kill it, if we are able to kill it at all . . . by burying it alive under a heap of facts . . . . [We] should not wrangle too much with incredulous outsiders about the conclusiveness of any one [study], but trust to the mass of evidence for conviction . . . . We must drive the objector into the position of being forced either to admit the phenomena as inexplicable, at least by him, or to

accuse the investigators either of lying or cheating or of a
blindness or forgetfulness incompatible with any intellectual
condition except absolute idiocy.[5]

Sidgwick was correct in the sense that objectors still accuse
psi investigators of lying or cheating or of "absolute idiocy." But
he was wrong in thinking that incredulity might be squashed by
a heap of facts. From Sidgwick's time to today, the facts about
psi have grown from a few morsels to a copious smorgasbord.
And it hasn't swayed academic opinion very much. Of more
than 3,000 traditional colleges and universities worldwide, less
than 1% host faculty who publicly admit interest in psi research.
By contrast, the great majority of psychology departments
house faculty who are interested in such specialized minutiae
that the average person wouldn't even recognize those topics as
belonging to the study of mind or behavior.

AN ANALOGY

My sense of this impasse is that psi is simply a phenomenon
that's ahead of its time, and science is slowly catching up to it.
There's a parallel with reports of *ball lightning*: a glowing, free-
floating, basketball-sized plasma that can persist for seconds to
minutes. Sightings of ball lightning are often associated with
storms, but it can also appear in clear weather. It can enter
buildings and squeeze through spaces smaller than its apparent
diameter; it floats slowly or zips quickly through the air; it is
said to emit little heat; and it can appear in a variety of colors
and brightness. These plasma balls have been observed to
hover and rotate, and roll on and bounce off surfaces. They disappear either by exploding or just vanishing quietly. Observations of ball lightning has been reported for centuries and
appear in the scientific literature starting in the mid-nineteenth
century.[6] As physicist D. J. Turner describes,

Nearly all the characteristic properties of ball lightning had been identified by the 1920s but, as a set, they have remained difficult to reconcile with the known laws of physics. Most attempts claiming a complete explanation of the phenomenon suggest that the authors ignored certain of the reported observations or refused to accept that they were attributable to ball lightning. Consequently, since at least the time of Arago [in 1855], many other scientists have been sceptical over the very existence of ball lightning as a physical entity. [7]

It took 100 years after the first scientific reports for the disciplines of chemistry and physics to advance into the new discipline of electrochemistry, and then another half century to advance to the point where plausible models of ball lightning could begin to be proposed. Only recently have fledgling demonstrations of ball lightning been produced in the laboratory. Since the turn of the twenty-first century, as observational and experimental evidence continue to accumulate, theoretical efforts are being proposed with greater frequency. One such report appearing in *Nature* suggested that ball lightning is caused by microscopic particles in the soil that absorb the energy from lightning strikes, float up, and slowly oxidize, releasing light and heat.[8] Another model proposed that the plasma ball is a thermochemical heat pump powered by the electric field of a thunderstorm.[9]

The point is that phenomena that are not easily accommodated by prevailing scientific theories are ignored and dismissed as impossible. If the phenomenon directly challenges basic assumptions, then they also attract ridicule. The same is true for psi. At some point in the future a new discipline will evolve. Within that discipline, models will arise that provide increasingly plausible explanations for psi experiences. By then the experimental evidence will also have advanced to the point where credible demonstrations can be repeated more easily. Like ball lightning, the phenomenon may be exquisitely sensitive and difficult to produce on demand, but it will appear often enough for the theories and observations to be put to the test.

THE SKEPTICS

We can easily forgive a child who is afraid of the dark; the real tragedy of life is when men are afraid of the light.

–Plato

I am often asked by journalists how to account for the fact that some scientists claim there's repeatable evidence for psi, while others don't. Both sides of the debate appear to be intelligent and knowledgeable. Both are aware of the strengths and limitations of meta-analyses. Is a resolution possible, or is this a case of permanently irreconcilable differences?

One way to answer this question is with a simplified political analogy: The "proponents" are liberals and the "skeptics" are conservatives. Both are interested in the same goal—to understand Nature. Both also wish to avoid mistakes that can be made in the pursuit of that goal. Scientific conservatives can't stand the idea of adulterating known truths by including potentially false ideas. And scientific liberals can't stand the idea that truth might be constrained by excluding potential new truths.

My prejudice is that it's more important to promote the serious study of novel ideas than it is to worry that some of those ideas might be wrong. I feel this way because history shows that virtually all exciting breakthroughs in science come from entertaining "crazy" ideas. Radically novel ideas *always* appear to be crazy at first, but genuinely crazy ideas do not last for long in the cold light of scientific scrutiny. So I believe there is sound justification to devote serious resources to their investigation. Of course, that's just my opinion. Other scientists prefer to protect the tried and true; they're uncomfortable with unorthodox ideas and would prefer to exclude anything that doesn't seem to fit with existing concepts.

DEBUNKING SKEPTICAL MYTHS

Another type of skepticism persists because assertions are repeated so often that through sheer repetition they start to take on an aura of truth. To help debunk this folklore, here is a myth-busting exercise based on recent critiques by York University psychologist James Alcock, who has widely published his skeptical views of parapsychology, and my answers to his critiques.[10]

Myth: Parapsychology is a pseudoscience. It claims to be like other scientific disciplines, but it has no core knowledge base, no set of constructs, no set of standard methodologies, and no set of accepted or demonstrable phenomena that all psi researchers would accept.

Fact: In 1969, parapsychology was accepted as an affiliate of the American Association for the Advancement of Science (AAAS), the largest scientific organization in the world and the publisher of *Science,* one of the top-ranked scientific journals. By inclusion in the AAAS, the Parapsychological Association is demonstrably a *bona fide* scientific discipline. By comparison, not one of the "professional" skeptical organizations, some of which even claim to be engaged in scientific investigation, is an affiliate of the AAAS. Assertions about a lack of core knowledge, constructs, and so on imply that to be scientific, members of a discipline must all agree upon a set of uniform beliefs. That's a quaint view of how science works. Pick up practically any scientific or scholarly journal and you'll quickly find that researchers are always engaged in vigorous debates and controversies. The moment a discipline collapses into a single set of beliefs, constructs, or even methods, it's no longer science, it's religion. As for "standard methods," many of them are described in this book.

Myth: Psi is unlike any other phenomenon studied in science. It's defined not in terms of what it *is,* but in terms of what it *isn't.*

Fact: In physics, when a charged particle is shot through a bubble chamber filled with liquid hydrogen, the particle is revealed by what it *isn't*—the stream of tiny bubbles that displace the liquid hydrogen. Similarly, the "what psi isn't" definition reflects how psi is investigated in the laboratory, not what it's thought to be. In other words, this question confuses the method of detection with the phenomenon itself. As a positive definition, psi is a means by which information can be gained from a distance without the use of the ordinary senses.

Myth: In mainstream science, one doesn't set out to evoke anomalies in the laboratory. They present themselves in the course of ordinary research and then science attempts to explain them.

Fact: This criticism occurs because the forest is overlooked while focusing on the trees. Psi research is not engaged in a quest for anomalies. It's engaged in investigating puzzling human experiences—often profoundly meaningful, and sometimes transformational, experiences—as reported by countless people throughout history.

Myth: The concept of a repeatable experiment in science means that any researcher with the proper expertise and equipment should be able to reproduce the reported results, not just those who are believers or enthusiasts. Parapsychology has never been able to produce a successful experiment that neutral scientists with the appropriate skill, knowledge, and equipment can replicate.

Fact: The meta-analyses discussed in previous chapters falsify this assertion, as does the more interesting question of what happens when skeptics try to repeat claimed effects. There are only a handful of examples. Consider the case of Stanley Jeffers, a skeptical physicist from York University. In 1992, Jeffers tried to repeat PK experiments similar to those reported by the Princeton Engineering Anomalies Research (PEAR) Laboratory.[11] He wasn't successful.[12] His skepticism was fueled by another PK study he reported in 1998, which also failed.[13] Then, in 2003 Jeffers coauthored a third study in which he finally reported a repeatable, significant PK effect.[14] So, can skeptics produce successful experiments? Yes, they can. They just hardly ever try.

Myth: The "experimenter effect," which asserts that some people can get significant psi results but others can't, is suspicious. It's just a hand-waving excuse used to explain the lack of consistency in experimental results.

Fact: Why do you suppose surgeons, trial lawyers, and business managers have different rates of success in their jobs? These folks are all talented, all highly trained, and all motivated to maintain fulfilling and successful careers. So what's the source of their differences? The answer is that people involved in jobs requiring authoritative interaction with others convey enormous influence whether they mean to or not. Their voice quality, posture, dress, perceived confidence, and mannerisms all play a significant role in how people respond to them.

But it's more than that. Their unstated *expectations* are also unconsciously conveyed to the experimental participants. In the 1950s, Harvard University psychologist Robert Rosenthal pioneered the study of the "experimenter expectancy effect," also known as the "interpersonal expectancy effect" and the Pygmalion effect. The latter refers to a Greek myth about a sculptor, Pygmalion, who carved a statue of a beautiful woman out of ivory. He fell so deeply in love with the statue that she (through the graces of the goddess Athena) came to life. This myth reflects the concept of a self-fulfilling prophecy.

When Rosenthal first proposed that experimenters' expectancies can be conveyed subtly to the participants in an experiment so as to create a self-fulfilling prophesy, the idea was considered laughable by some and a revolutionary advancement by others. As Rosenthal put it,

That this research was received with ambivalence is illustrated by the receipt of two letters on the same day: The first letter rejected the paper for publication in a prestigious journal, and the second letter announced that the paper had received the Socio-Psychological Prize for 1960 from the American Association for the Advancement of Science.[15]

This concept has since been studied in hundreds of experiments with teachers, attorneys, judges, business managers, and health care providers. It has been repeatedly shown that expectations unintentionally affect the responses of research participants, pupils, jurors, employees, and patients.[16] Rosenthal and others have shown that these effects are not mere subtleties; they have meaningful consequences in the real world. For example, in medical contexts the degree to which a physician is effective in persuading their patients to enter into treatment can be predicted from "their tone of voice in talking to or about their patients."[17] Another study found that "surgeons who used a bossy tone of voice when talking to their patients were more likely to be sued by their patients than were surgeons who used a more respectful tone."[18]

From this perspective, it would be suspicious if all psi investigators were uniformly successful. The idea that any suitably trained experimenter should be able to produce a successful result in any experimental context is a nice ideal, but it's unrealistic. One of the factors in models proposed to explain the Pygmalion effect is called "climate." This refers to whether the investigator has a warm, inviting, permissive interpersonal style vs. a cold, aloof, pessimistic attitude. Is it any wonder that a suspicious or cynical attitude, which unconsciously permeates the interpersonal mannerisms of many skeptics, fails to get results, while warm and enthusiastic proponents do? These factors aren't the only reasons for success vs. failure, but they do play an important role.

What skeptics have in mind when offering this critique are extremely stable effects like gravity. Gravity doesn't care if one is skeptical or not. But this is never the case when it comes to evaluating human performance, even exceptionally talented human performance. Consider the "home team advantage" in sports.[19] As reported in the *Journal of Economics and Business,*

> In the National Football League, home teams won 58% of games over the period 1981–1996. . . . However, for a subset of games which had national focus (i.e., Monday night and playoff games), betting on the home team produced a [59.2%] win rate . . . and betting on underdog Monday night and layoff home teams produced a [65.5%] win rate. These results suggest that the home field advantages recognized in the sports psychology literature are increased under the public attention of national exposure to a larger extent than is recognized by bettors.[20]

If highly skilled sports performance can differ dramatically depending on who's watching even among professionals at the top of their game, why should we expect any less in psi experiments?

Myth: Parapsychologists cannot make predictions before running experiments and then confirm them.

Fact: If this means knowing in advance exactly which conditions will produce a successful outcome with 100% certainty, then the criticism is true. But of course, that level of absolute certainty doesn't exist for any form of human performance, so the requirement is unrealistic. Studies like the ganzfeld telepathy experiment are designed with clear predictions, and the cumulative data show that those predictions are confirmed to very high levels of confidence.

Myth: Parapsychology uses statistics not to evaluate the effect of one variable on another, but as a way to infer the presence of psi itself. That's not a legitimate way of demonstrating an effect.

Fact: Burton Camp, president of the Institute of Mathematical Statistics, settled this question in favor of parapsychology as early as 1937. Regarding the statistical methods used by J. B. Rhine to infer the presence of psi in ESP card tests, Camp wrote:

> Dr. Rhine's investigations have two aspects: experimental and statistical. On the experimental side mathematicians, of course, have nothing to say. On the statistical side, however, recent mathematical work has established the fact that, assuming that the experiments have been properly performed, the statistical analysis is essentially valid. If the Rhine investigation is to be fairly attacked, it must be on other than mathematical grounds.[21]

Myth: Theorizing in the absence of reliable data, especially when it attempts to interpret quantum mechanical theory in such a way as to accommodate psi, lends an unjustified patina of scientific respectability to parapsychology, especially in the eyes of those who are outside the world of physics.

Fact: "Reliable" data is always in the eyes of the beholder. For those unwilling to accept the data, then of course theorizing is premature. But for those who have accepted the data, a more comprehensive explanation is demanded. And, as discussed in previous chapters, I believe that the connection between modern physics and psi phenomena goes far beyond a mere patina.

Myth: Parapsychology fails to jibe with other areas of science. If parapsychology is right, then physics and biology and neuroscience are horribly wrong in some fundamental respects.

Fact: Such comments might be expected from a theist who felt his or her faith were being threatened by new facts, but they seem utterly out of place in a scientist. Obviously existing scientific knowledge is not "horribly wrong." But neither is it absolutely correct. Quantum physics is radically different from classical physics, and yet it encompasses the former view. The same is true for advancements in all scientific disciplines. To assume otherwise is a failure of imagination and a denial of history.

Myth: Supposed psi influences, unlike any known energy, are invariant over distance. Time produces no barrier either, for such influences are said to be able to operate backwards and forwards in time.

Fact: This argument would have been a showstopper in the seventeenth century, but it overlooks progress in physics since then. Spacetime nonlocality in quantum physics and time-symmetry in the formalisms of classical and quantum mechanics are well-established. Also, while the evidence suggests that psi is not strongly bound by the constraints of space and time, it is not yet certain that psi is absolutely space-time independent.

Myth: The vast majority of academic psychologists assume that psi phenomena have never been shown to exist. If compelling evidence for the reality of psi appeared some day, then psychologists wishing to explore an exciting new area of research would knock over parapsychologists in the stampede.

Fact: Academic psychologists tend to avoid psi phenomena because of severe distortions in how this topic is portrayed in textbooks. A review of introductory psychology textbooks in 2002 showed that only 33 of 57 popular texts (58%) mentioned psi at all, and those that did devoted an average of only 2.4 pages to the topic.[22] Rhine's ESP cards were discussed in 14 books, and the ganzfeld telepathy tests were discussed in 24 books. That's all. There is not one word about the hundreds of other experiments we've reviewed here, which are themselves a subset of the larger literature. Of all individual publications mentioned, the highest number, a total of 63, referred to discussions of the ganzfeld tests described in the journal *Psychological Bulletin*. But the second highest, a stunning 58 citations, referred to articles in the magazine, *Skeptical Inquirer*. This should make your hair stand up. It's like trying to sustain a serious scientific discussion based on citations from tabloids with such sober stories as "Jesus' sandal found in Central Park!" and "Scientist reveals thunder caused by fat people doing jumping jacks!"[23] If this is the type of scholarly information being fed to impressionable psychology students, it's not surprising that whole generations of future academic psychologists assume there's nothing to it. Also, after the 1994 publication of the successful ganzfeld experiments in *Psychological Bulletin*, no replication attempts were subsequently reported in peer-reviewed journals by the supposed stampede of academic psychologists. Not one.

Myth: If psi were real, someone would have already won one of the many prizes offered by skeptics to demonstrate existence of psychic phenomena.

Fact: It's true that conjurers masquerading as psychics won't win those prizes, or by people who think they're more psychic than they really are. But I doubt that such prizes, if genuine, will stand forever. This is because phenomena like psi that have withstood the test of time, and the scrutiny of science, are very likely to be real. As science advances, those phenomena will emerge out of the uncertain realm of the paranormal and into the more comfortable domain of the normal. Such transitional phenomena are called the "perinormal," meaning near-normal, by fiercely skeptical Cambridge University zoologist Richard Dawkins. Dawkins offered this new term at a skeptic's conference in January 2005. As author Ted Dace wrote in reviewing that conference,[24] when Dawkins was asked about one of the prizes offered for demonstrations of the paranormal, he responded: "About the million dollar prize, I would be worried if I were you because of the fact that we have perinormal possibilities." It appears that among skeptics, the smart players are beginning to hedge their bets.

Beyond betting, one wonders if these prizes are really worth the effort. For an individual who is actually able to demonstrate

strong, reliable psi effects under the harsh scrutiny of determined skeptics (this would be an exceptionally rare person), it
might be worth it as his or her costs might be minimal. But for
the types of psi effects observed in the laboratory, even a million
dollar prize wouldn't cover the costs of conducting the required
experiment. Assuming we'd need to show odds against chance
of say 100 million to 1 to win a million dollar prize, when you
calculate how many repeated trials, selected participants, multiple experimenters, and skeptical observers are necessary to
achieve this outcome, the combined costs turn out to be more
than the prize. So, from a purely pragmatic perspective, the various prizes offered so far aren't sufficiently enticing.

Fact: Statistical analyses are being used to define and defend
the importance of differences in psi experiments that are so
small that they would have carried no interest to researchers of
a century ago.

Fact: The charge on the electron is also very small. But so
what? This confuses the magnitude of an effect with its existence. In any case, many modern psi experiments test randomly
selected college sophomores and other unselected participants.
As a result, the weak effects often observed in the laboratory are
probably due to the fact that the people being tested aren't talented in the skills of interest.

Say we were interested in studying high jumping. We've
heard tales of people jumping as high as 6 feet, but we don't believe it. It seems to defy the laws of gravity to jump higher than
one's own height. But we're willing to put the claim to the test.
So we select 100 college sophomores at random, we measure
how high each can jump, and from this we build up a distribution of possible jumping heights. We find that the average student can jump 3.4 feet. We compare this to the claim of 6 feet,
and we prove to our satisfaction that such claims are nonsense.
Not one sophomore comes close to jumping that high. And yet,
the world's high jump record is not merely 6 feet, but over 8
feet.[25] Only exceptionally talented athletes can approach that

height, so if we relied exclusively on unselected volunteers we'd never be able to confirm the exceptional claims.

Some people selected for psi talent, and tested over long periods of time, have shown comparatively strong, reliable effects. Such people are rare, but they do exist. One example is Joseph McMoneagle, remote viewer #001 in the U.S. Army's formerly Top Secret project codenamed GRILLFLAME, STARGATE, and other exotic names.[26] McMoneagle has been repeatedly tested in numerous double-blind laboratory experiments and has been shown to have an ability to describe objects and events at a distance and in the future, sometimes in spectacular detail. In one experiment, all that McMoneagle knew was that a person he hadn't met before would be visiting a technological target, at a certain time, somewhere that could be reached within an hour's drive around Silicon Valley in Northern California. The number and range of possible technological targets that one can get to in a short drive around Silicon Valley is gigantic. As it turned out, the target that the person arrived at was a particle beam accelerator, and that's what McMoneagle drew (Figure 14–1).[27]

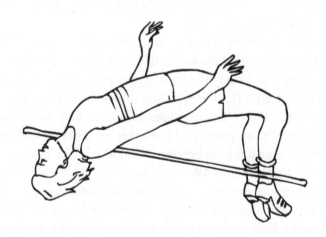

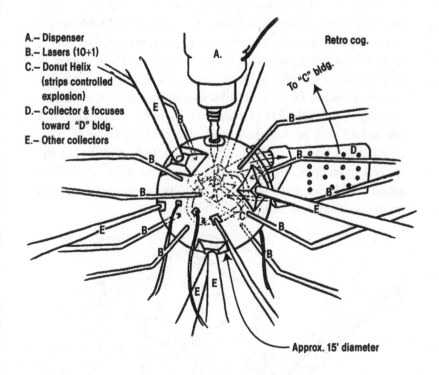

A.– Dispenser
B.– Lasers (10+1)
C.– Donut Helix
(strips controlled
explosion)
D.– Collector & focuses
toward "D" bldg.
E.– Other collectors

Retro cog.

To "C" bldg.

Approx. 15' diameter

Figure 14–1. Drawing of a distant technological target by military remote viewer, Joseph McMoneagle. As it turned out, this is a good representation of that target.

TOWARDS PSI APPLICATIONS

Given the (albeit rare) availability of such skills, I became interested in seeing whether expert psi might be useful in helping to invent practical devices. For some years I've been interested in creating a "psi-switch," a technological way of detecting mental intention at a distance. A U.S. patent (number 5830064) based on the research of the Princeton Engineering Anomalies Research Laboratory was granted for a psi-based effect on November 3, 1998. So the concept of psi-based technologies may not be as improbable as some believe.

In one of my psi-switch projects, over the course of a year and in a dozen different sessions, I wondered if McMoneagle could explore the near-term future to describe the first prototypes of operating psi-switch devices. I figured that if it were at all possible to describe a future invention, I didn't want information from too far in the future, as that would be like demonstrating a portable DVD player to Benjamin Franklin and asking him how he thought it worked. He'd have no hope of explaining it. Likewise, a far futuristic psi-amplification device described in today's terms wouldn't make any sense at all. So I asked McMoneagle to provide a glimpse of prototype devices on the near horizon. And he did (Figure 14–2).

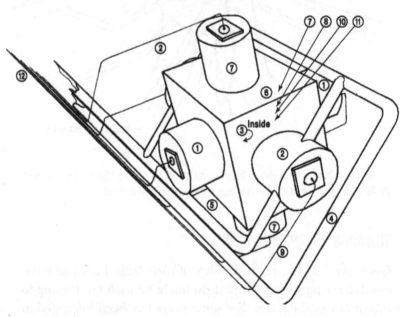

Figure 14-2. Sketch of a future intention-amplifying, mind-operated "psi switch," drawn by remote viewer Joseph McMoneagle. This drawing is reproduced here only as a teaser to illustrate that sometimes remarkably detailed technical information can be obtained with talented psychics.

The topic of possible psi applications is intriguing but tricky, and it deserves a careful treatment beyond the scope of this book.[28] For now, extrapolating from current knowledge and trends, I speculate that in the future we may see psi at the core of exotic forms of communication and prediction technologies. There's already evidence that psi effects can be amplified through the use of statistical error-correction techniques.[29] These applications will probably not become part of the world of consumer electronics in the short run, except perhaps in the form of simple toys or games. We're likely to see an increasing number of techniques and training programs developed for augmenting intuition, as well as methods for detecting intention at a distance. And we'll almost certainly see growing interest in "collective psi," effects that become noticeable in the behavior and decisions of groups rather than individuals.[30] As the field consciousness experiments, Global Consciousness Project, and online psi tests are beginning to show, there's much to be learned from observing our collective intentions and intuitions. These effects may lead to whole new classes of "social psi" applications.

CONCLUSION

Some day psi research will be taught in universities with the same aplomb as today's elementary economics and biology. It will no longer be considered controversial, but just another facet of Nature one learns as part of a well-rounded education. In that future no one will remember that psi was once considered the far fringe of science. We'll argue over new controversies not even imagined yet.

History shows that as the scientific frontiers continue to expand, the supernatural evolves into paranormal, and then into normal. During the transitional periods there is much gnashing of teeth. But with determination and courage, progress is relentless.

Dennis and Terrence McKenna offered one of my favorite thoughts about the process of scientific discovery: *As the bonfires of knowledge grow brighter, the more the darkness is revealed to our startled eyes.* This evokes an image of sitting in pitch blackness around a small campfire, in a very dark forest, on a moonless, overcast night. The first tiny spark of understanding illuminates just our frightened faces. As the fire grows, we begin to see each other more clearly. Then we begin to see the campsite itself, and later some trees. At each stage of expansion, as our knowledge about who and what we are advances, the fire glows brighter. But with each new expansion, we also begin to grasp just how *large* the space is that we've been huddled in, and how immense the still-dark space is that's yet to be explored. What new wonders lie hidden in those shadows?

The spark of psi is glowing brighter than ever before, but enlightenment is fleeting and fragile. It's risky exploring unnamed realms. Those who are frightened of the darkness often refuse to look, and they don't want anyone else to look either. But what's life worth living for? Pushing the horizons of science invariably involves risk and controversy, but the potential of real discovery always makes those risks worthwhile. Be bold. Cultivate that spark of curiosity. Courage!

ACKNOWLEDGMENTS

I am indebted to the members of the Institute of Noetic Sciences (IONS) for their support of a research organization dedicated to exploring the frontiers of human consciousness. I am especially thankful for the support of President James O'Dea, Vice President Marilyn Schlitz, IONS founder, and Apollo 14 astronaut Edgar Mitchell, and the generous support of Richard Adams, Peter Baumann, Michael Breland, Charles Brush, Sandra Hobson, Claire Russell, William Froelich, George Zimmer, Luis Portela and the Bial Foundation, and the Fetzer Institute. I'd also like to acknowledge the memories of Wink Franklin, former IONS president who was so supportive of my work, and staff member Violet Cleveland.

I'm indebted to many friends and colleagues for valuable discussions that have helped to clarify and refine my thoughts. I especially thank Carlos Alvarado, Dick Bierman, Damien Broderick, Colin Cherot, Larry Dossey, Thomas Etter, James Fallon, Nick Herbert, Amy Lansky, Lynne Mason, Edwin May, Garret Moddel, Michael Murphy, Robert McConnell, Roger Nelson, Lou Rudolph, Stephan Schwartz, Rupert Sheldrake, Richard Shoup, Henry Stapp, Russell Targ, Caroline Watt, and Richard Wiseman. Any mistakes of fact or interpretation in this book are my fault, not theirs.

I am grateful to Sandra Martin of Paraview Pocket Books and Lisa Hagan of the Paraview Literary Agency for guiding the manuscript through the labyrinth of the publishing business, and to Patrick Huyghe for his insightful editorial supervision.

I thank Jenny Matthews, Charlene Farrell, Cassandra Vieten, Tina Amorok, Kelly Durkin, and Gail Hayssen of the IONS Research Department for providing a work environment where the serious consideration of controversial ideas is not only encouraged but actually allowed to thrive. Such genuinely supportive environments are rare to find, even among facilities supposedly engaged in innovative research.

I'd like to acknowledge Silicon Valley luminaries David Liddle, Federico Faggin, Andrew Singer, Denise Caruso, and Richard Shoup for supporting and encouraging my research.

I am indebted to the hordes of children in my neighborhood for screaming at full volume while playing in the street outside my home office, for, as P. G. Wodehouse once said, without their assistance this book would have been finished in half the time.

I thank my wife Susie and two small dogs named Wilbur and Bunny for their good humor, occasional barking (the dogs, not Susie), and gentle patience while I was writing. I thank my parents, Hilda and Jerry, for teaching me the joys of learning. And finally, I'd like to say hello to my big brother Len, dentist and drama teacher extraordinaire. Hi, Len.

IMAGE ACKNOWLEDGMENTS

Figure 6–8 reprinted from Sanchez, J., Dohoo, I., Carrier, J., DesCôteaux, L. (2004). A meta-analysis of the milk-production response after anthelmintic treatment in naturally infected adult dairy cows. *Preventive Veterinary Medicine,* **63**, (3–4), 237–56, with permission from Elsevier publishers.

Figure 6–9 reprinted from the Particle Data Group's document located at http://pdg.lbl.gov/2005/reviews/historyrpp.pdf as of April 2005, with permission from the Particle Data Group.

Figure 7–2 reprinted with permission from Schmidt, S., Schneider, R., Utts, J., Walach, H. (2004). Distant intentionality and the feeling of being stared at: Two meta-analyses. *British Journal of Psychology,* **95**, 235–247. © The British Psychological Society.

Figure 7–3 reprinted from Standish, L. J., Johnson, L. C., Richards, T. and Kozak, L. (2003). Evidence of correlated functional MRI signals between distant human brains. *Alternative Therapies in Health and Medicine,* **9**, 122–128, with permission from InnoVision Communications in the format Trade Book via Copyright Clearance Center.

Figure 7–4 reprinted from Kittenis, Caryl and Stevens. "Distant psychophysiological interaction effects between related and unrelated participants," *Proceedings of the 2004 Parapsychological Association,* with the kind permission of Mario Kittenis.

Figure 9–3 reprinted from Jahn, R. G., Dunne, B. J., Nelson, R. D., Dobyns, Y. H., and Bradish, G. J. (1997). Correlations of random binary sequences with pre-stated operator intention: A review of a 12-year program. *Journal of Scientific Exploration,* **11** (3), 345–367, with permission from the *Journal of Scientific Exploration.*

Figure 10–4 reprinted from McCraty, R. Atkinson, M., and Bradley, R. T. (2004). Electrophysiological evidence of intuition: Part 1. The Surprising role of the heart. *Journal of Alternative and Complementary Medicine,* **10**, 133–143, with permission from the *Journal of Alternative and Complementary Medicine.*

Figure 10–5 reprinted from Spottiswoode, S. J. P., and May, E. C. (2003). Skin conductance prestimulus response: analyses, artifacts and a pilot study. *Journal of Scientific Exploration,* **17** (4), 617.641, with the kind permission of James Spottiswoode.

Figure 10–6 reprinted from Bierman, D. J. and Scholte, H. S. (2002). Anomalous anticipatory brain activation preceding exposure of emotional and neutral pictures. Paper presented at Toward a Science of Consciousness, Tucson IV, with the kind permission of Dick Bierman.

Figure 11–1 reprinted from Radin, D. I., Taft, R. and Yount, G, (2004). Possible effects of healing intention on cell cultures and truly random events. *Journal of Alternative and Complementary Medicine,* **10**, 103–112, with permission from the *Journal of Alternative and Complementary Medicine.*

Figures 11–7 and 11–8 reprinted from the Global Consciousness Project Web site, http://noosphere.princeton.edu, with the kind permission of Roger Nelson.

Figure 11–11 reprinted Nelson, R. D., Radin, D. I., Shoup, R. and Bancel, P. (2002). Correlations of continuous random data with major world events. *Foundations of Physics Letters,* December 2002, vol. 15, no. 6, pp. 537–550 (14), with permission from Springer Science and Business Media.

Clipart images © 2005 JupiterImages Corporation (ClipArt.com).

ENDNOTES

Chapter 1: In the beginning

1. Manx is the traditional language of the Isle of Man, off the coast of Scotland, where the word for psychic is *sheekagh*.
2. Kempner, J., Perlis, C. S., and Merz, J. F. (2005). Forbidden knowledge. *Science*, **307**, 854.
3. Lindorff, D. (2004). *Pauli and Jung: The Meeting of Two Great Minds.* Wheaton, IL: Quest Books.
4. Psi might be a side effect of the universe looking at itself.
5. Bell Laboratories is where the transistor and much of the infrastructure for local, national, and global telecommunications were invented. Bell Labs is now part of Lucent Technologies.
6. The AAAS is the world's largest general scientific society, serving over 200 affiliated societies and academies of science and some 10 million individual members.
7. Some conspiracy theorists assert that this program was part of the CIA's nefarious MKULTRA "mind-control" projects of the 1950s and 60s. That belief is mistaken. STARGATE and related projects at SRI International had nothing to do with MKULTRA.
8. Kirshner, R. P. (2003). Throwing light on dark energy. *Science*, **300** (5627), 1914–1918.
9. Clark, S. (March 12, 2005). Did we miss dark energy first time around? *New Scientist*, p. 14.
10. Bejerano, G., Pheasant, M., Makunin, I., Stephen, S., Kent, W. J., Mattick, J. S., and Haussler, D. (2004). Ultraconserved elements in the human genome. *Science*, **304** (5675), 1321–1325.
11. Shankle, W. R., Rafii, M. S., Landing, B. H., and Fallon, J. H. (1999). Approximate doubling of the numbers of neurons in the postnatal human cortex and in 35 specific cytoarchitectonic areas from birth to 72 months. *Pediatric and Developmental Pathology*, **2**, 244–259.
12. The student's cortex was compressed into a thickness of less than a

millimeter due to hydrocephalus. Lewin, R. (1980). Is your brain really necessary? *Science,* **210** (1232).

13. Seife, C. (2004). Outlook for cold fusion is still chilly. *Science,* **306** (5703), 1873. From this article: "Several reviewers were indeed extremely critical of the research, saying that many of the experiments were poorly conducted, had results that were inconsistent with each other, and often weren't reproducible . . . . [But] at the same time, about one-third of the reviewers . . . were receptive to claims of cold fusion. 'There is strong evidence of nuclear reactions in palladium,' one wrote. Said another: 'Further work that would add to the understanding of [low-energy nuclear reactions] is warranted and should be funded by U.S. funding agencies.'"

14. Bekenstein, J. D. (February, 2004). Information in the holographic universe. *Scientific American* (online), 29–36.

15. E.g., Ghosh, S., Rosenbaum, T. F., Aeppli, G., and Coppersmith, S. N. (2003). Entangled quantum state of magnetic dipoles. *Nature,* **425**, 48–51; Sackett, C. A., Kielpinski, D., King, B. E., Langer, C., Meyer, V., Myatt, C. J., Rowe, M., Turchette, Q. A., Itano, W. M., Wineland, D. J., and Monroe, C. (2000). Experimental entanglement of four particles. *Nature,* **404**, 256–259; Blinov, B., Moehring, D. L., Duan, L. M., and Monroe, C. (2004). Observation of entanglement between a single trapped atom and a single photon. *Nature,* **428**, 153–157; Julsgaard, B., Kozhekin, A., and Polzik, E. S. (2001). Experimental long-lived entanglement of two macroscopic objects. *Nature,* **413**, 400–403; Zhao, Z., Chen, Y.-A., Zhang, A.-N., Yang, T., Briegel, H. J., and Pan, J.-W. (2004). Experimental demonstration of five-photon entanglement and open-destination teleportation. *Nature,* **430**, 54–58; Sørensen, A., Duan, L.-M., Cirac, J. I., and Zoller, P. (2001). Many-particle entanglement with Bose–Einstein condensates. *Nature,* **409**, 63–66; Kwiat, P. G., Barraza-Lopez, S., Stefanov, A., and Gisin, N. (2001). Experimental entanglement distillation and "hidden" non-locality. *Nature,* **409**, 1014–1017; Marcikic, I., de Riedmatten, H., Tittel, W., Zbinden, H., and Gisin, N. (2003). Long-distance teleportation of qubits at telecommunication wavelengths. *Nature,* **421**, 509–513.

16. Mermin, N. D. (April 1985). Is the moon there when nobody looks? Reality and the quantum theory. *Physics Today,* 45.

17. Johnson, G. (July 16, 2001). Refining the cat's-eye view of the cosmos. *New York Times.* "That spooky connection between tiny particles is appearing everywhere, and its consequences are even affecting the world that we experience."

18. Brooks, M. (March 27, 2004). The weirdest link. *New Scientist.*

19. http://inexistant.net/Gilles/en/ as of December 29, 2004.

20. Hsieh, J.-Y., Li, C.-M., and Chuu, D.-S. (2004) A simplification of entanglement purification. *Physics Letters A, 328, 94–101.*

21. Brooks, M. (2004), op. cit., also endnote 15.

22. Brooks, M. (2004), op. cit., also endnote 15.

23. Buchanan, M. (March 12, 2005). Simple twist untangles quantum computing. *New Scientist,* p. 9.

24. Arndt, M. and Zeilinger, A. (March 2005). Probing the limits of the quantum world. *Physics World.* Online at http://www.physicsweb.org/articles/world/18/3/5/1.

25. Physicist Evan Harris Walker was one of the first to suggest this. Also see Josephson, B. D. and Pallikari-Viras, F. (1991). Biological utilisation of quantum non locality, *Foundations of Physics,* **21**, 197–207. Also, Hammeroff, S. and Penrose, S. (1996). Orchestrated reduction of quantum coherence in brain microtubules: A model for consciousness? In Hameroff, S. R., Kaszniak, A. W., and Scott, A. C. (Eds.), *Toward a Science of Consciousness,* Cambridge, MA: MIT Press, pp. 507–540; Hammeroff, S. (1998). Funda-Mentality: Is the conscious mind subtly linked to a basic level of the universe? *Trends in Cognitive Sciences,* **2** (4), 119–127.

26. Summhammer, J. (March 15, 2005). Quantum cooperation of insects. ArXiv: quant-ph/0503136 v1. Online at http://xxx.arxiv.cornell.edu/PS_cache/quant-ph/pdf/0503/0503136.pdf

27. Brooks, M. (2004), op. cit.

28. Some scientists don't accept the nonlocal implications of entanglement because to them it sounds like magic. But that group is a fast-shrinking minority. Entanglement can be demonstrated now to very high levels of certainty in the physics laboratory, so the experimental fact is now independent of quantum theory. Notice the parallel with psi: Some scientists reject psi because it looks like magic. And yet repeated experiments show to high levels of certainty that these phenomena exist.

29. Targ, R. (2004). *Limitless Mind.* Novato, CA: New World Library. p.5; Playfair, G. L. (2003). *Twin Telepathy: The Psychic Connection.* London: Vega.

30. The phrase "Entangled Brains" was used as a title for a letter to the editor in *New Scientist* magazine on May 1, 2004. The writer was pointing out the quantum-psi similarities discussed in this book.

31. The original article cited the U.S. National Institutes of Health as the

source of funding for the study, but the Freedom of Information Act eventually revealed that the actual source of funds was the Central Intelligence Agency. This unfortunately fuels the fears of those who assume there are nefarious connections between psi and shadowy government programs. Based on my understanding of psi, I believe that manipulation of mass media would be a far more effective social engineering tool than the use of any form of psi.

32. Wackermann, J. (2004). Dyadic correlations between brain functional states: Present facts and future perspectives. *Mind and Matter*, **2** (1), 105–122.

Chapter 2: Naked Psi

1. Hans Berger was raised in Coburg, Germany.
2. Berger, H. 1940. *Psyche*. Jena: Gustav Fischer, p. 6.
3. Millett, D. (2001). Hans Berger: From psychic energy to the EEG. *Perspectives in Biology and Medicine*, **44** (4), 522–542; Bender, H. (1963). Hans Berger and an energetic theory of telepathy. *Zeitschrift für Parapsychologie und Grenzgebiete der Psychologie*, **6**, (2/3): 182–191.
4. This refers to the EEG correlation experiment published in *Science* in 1965, as discussed later.
5. These historical topics are discussed in Chapter 4.
6. Gabbey, A. and Hall, R. E. (1998). The melon and the dictionary: Reflections on Descartes's dreams. *Journal of the History of Ideas*.
7. From "How to build a universe that doesn't fall apart two days later," by Philip K. Dick, 1978.
8. This excerpt of Dr. MacGregor's story is reprinted with permission.
9. Feather, S. and Schmicker, M. (2005). *The Gift: ESP–The Extraordinary Experiences of Ordinary People*. New York: St. Martin's Press.
10. The end date is arbitrary, referring only to when I conducted this analysis, in July 2003.
11. To do this, I used a database of word associations developed by psycholinguists at the University of Edinburgh in the early 1970s, known as the Edinburgh Associative Thesaurus; see Kiss, G. R., Armstrong, C., Milroy, R., and Piper, J. (1973). An associative thesaurus of English and its computer analysis. In Aitken, A. J., Bailey, R. W., and Hamilton-Smith, N. (eds.), *The Computer and Literary Studies*. Edinburgh: University Press. URL: www.eat.rl.ac.uk as of January 2005. If a word was an exact match, a computer algorithm assigned it a score of 100, otherwise it was assigned a score provided by the associative database. For example, say the relationship bird→(airplane had

a word association value of 11. This means that when presented with the word "bird," on average 11 out of 100 people would respond with "airplane" as the first word that sprang to mind. The associative values for each word on each trial were added up, and then the sums for all trials provided on a given day were summed to create a single terrorism concept-matching score per day. Those values were in turn transformed into a standardized score to judge "how much" terrorism ideation appeared in the precognition trials on a daily basis. The average terrorism ideation association score per day $m = s/w$, where s is the grand sum of association scores per trial per day, and w is the number of words contributing to the formation of the grand sum in a given day. Then the skew of the distribution of the prior 30 days of m was formed, and those values were normalized.

12. These odds were determined via a nonparametric bootstrap analysis.

13. This curve is created by taking the observed sum of trials and hits obtained in a period of 30 days, from day i to day $i+29$, then creating a proportion $p_1 = $ hits/trials, and from that $z = (p_1 \text{-} p_0 / \sqrt{p_0 q_0 / N}$, where p_0 is the chance expected hit rate of 20%, $q_0 = 1 \text{-} p_0$, and N is the number of trials. Then we create a new 30-day z score by incrementing the day by 1, and repeating for all available data.

14. To calculate these probabilities, first we scrambled the order of the days, then we recalculated the sliding z scores as described in the endnote above. Then we find the minimum z value obtained in our curve, and also how many days this minimum is to 9/11. This process was repeated 10,000 times to build up a distribution of minimum z scores and number of days to 9/11. Then we compare the original z minimum and days to 9/11 against these bootstrap-generated distributions.

15. The remote viewing bootstrap result was associated with $z = \text{-}3.5$ and the card test result with $z = \text{-}3.4$. Combined this is associated with a (Stouffer) $z = \text{-}4.9$. This is an exploratory result and should be interpreted in that light.

Chapter 3: Who Believes?

1. Etter, T. (September 1997). Theories of psi, *ANPA* (Alternative Natural Philosophy Association) *West Journal*, 7 (1).

2. Melissa Pollak of the National Science Foundation reportedly described herself as "personally concerned with rampant misunderstanding of science in this country, in people's not knowing how science works and with belief in such phenomena as psychic powers. 'Not to spoil anyone's enjoyment of an entertaining television show [about a woman

who talks to the dead], but if you really believe that this person has psychic ability, then I see that as something society should be concerned about.'" Quoted from a February 18, 2005, news release from the AAAS about a seminar on the public's perception of science at the annual meeting of the AAAS in Washington, DC.

3. Pollak, M. F., (2002). Chapter 7. Science and technology: Public attitudes and public understanding, *Science and Technology Indicators: 2002*, National Science Foundation.

4. National Science Foundation, Division of Science Resources Statistics (NSF/SRS), NSF survey of public attitudes toward and understanding of science and technology, 2001.

5. Newport, F. and Strausberg, M. (2001). *Poll analyses: Americans' belief in psychic and paranormal phenomena is up over last decade.* Gallup Organization (Princeton, NJ).

6. URL as of January 2005: http://www.norc.uchicago.edu/projects/gensoc.asp; The National Opinion Research Center receives its core funding from the National Science Foundation.

7. Goode, E. (Jan.–Feb. 2002). Education, scientific knowledge, and belief in the paranormal. *Skeptical Inquirer.*

8. Rice, T. (2003). Believe it or not: Religious and other paranormal beliefs in the United States. *Journal for the Scientific Study of Religion,* **41** (1), 95–106.

9. The Harris Poll, February 26, 2003, http://www.harrisinteractive.com/harris_poll/index.asp?PID=359.

10. Newport, F., and M. Strausberg. (June 8, 2001). Americans' belief in psychic and paranormal phenomena is up over last decade. *Gallup News Service.* Available at http://www.gallup.com/poll/releases as of January 2005.

11. Sjödin, U. (2002). The Swedes and the paranormal. *Journal of Contemporary Religion,* **17** (1).

12. Roe, C. (1999). Critical thinking and belief in the paranormal: A reevaluation. *British Journal of Psychology,* **90**, 85–98.

13. Hughes, M., Behanna, R. and Signorella, M. (2001). Perceived accuracy of fortune telling and belief in the paranormal. *Journal of Social Psychology,* **141** (1), 159–160.

14. Wolfradt, U. (1997). Dissociative experiences, trait anxiety and paranormal beliefs. *Personality and Individual Differences,* **23** (1), 15–19.

15. Dag, I. (1999). The relationships among paranormal beliefs, locus of control and psychopathology in a Turkish college sample. *Personality and Individual Differences,* **26**, 723–737.

16. Goulding, A. (2005). Healthy schizotypy in a population of paranormal believers and experients. *Personality and Individual Differences,* **38**, 1069–1083.

17. Shermer, M. (2002). Smart people believe weird things. *Scientific American.*

18. Doctors at the time believed that the circulatory system carried both blood and spirits (air), and that the purpose of the heart was essentially to assist the lungs.

19. *Exercitatio Anatomica de Motu Cordis et Sanguinis in Animalibus,* URL: http://www.fordham.edu/halsall/mod/1628harvey-blood.html as of December 19, 2004.

20. Translation by Harald Walach, p. 67–68, in Samulei Institute for Information Biology's *Proceedings of a Workshop on Generalized Entanglement;* original from Parisano, E. (1647). *Recentiorum disceptationes de motu cordis, sanguinis et chili.* Leiden: Ioannis Maire.

21. Simons, D. J. and Chabris, C. F. (1999). Gorillas in our midst: sustained inattentional blindness for dynamic events. *Perception,* **28**, 1059–1074.

22. Our questions were based partially on the work of psychologist David Ritchey, as reported in his 2003 book, *The H.I.S.S. of the A.S.P.: Understanding the Anomalously Sensitive Person* (Terra Alta, WV: Headline Books). For those interested in more information on highly sensitive people, I recommend psychiatrist Vernon Neppe's Web site (www.pni.org) and psychologist Elaine Aron's excellent series of books on the "highly sensitive person" (www.hsperson.com).

23. Most of these differences were statistically significant, as indicated by the error bars.

24. The categories in this graph include electrical sensitivities, vivid imagination, empathize easily, have unusual experiences at 3 a.m., experience colors and sounds mixed together, have seen unexplained glowing lights, heard unexpected sounds, smelled strange odors, experienced extreme bliss, periods of missing time, are present when electronics break, notice that lights flash in your presence, are troubled by fluorescent lights, feel energy flows within the body, have experienced a life-changing event, find it easy to recapture events in memory, sense thunderstorms approaching, seem to cause lights to flicker, and are especially sensitive to the environment.

25. Persinger, M. A., Koren, S. A., and O'Connor, R. P. (2001). Geophysical variables and behavior: CIV. Power-frequency magnetic field transients (5 microtesla) and reports of haunt experiences within an

electronically dense house. *Perceptual and Motor Skills,* **92**, 673–674; Neppe, V. M. (1983). Temporal lobe symptomatology in subjective paranormal experients. *Journal of the American Society for Psychical Research,* **77** (1), 1–29.

26. Persinger, M.A., Tiller, S. G., and Koren, S. A. (2000). Experimental simulation of a haunt experience and paroxysmal electroencephalographic activity by transcerebral complex magnetic fields: Induction of a synthetic ghost? *Perceptual and Motor Skills,* **90**, 659–674.

27. Khamsi, R. (December 9, 2004). news@nature online, "Electrical brainstorms busted as source of ghosts."

28. Persinger, M. A., Roll, W. A., Tiller, S. G., Koren, S. A., and Cook. C. M. (2002). Remote viewing with the artist Ingo Swann: Neuropsychological profile, electroencephalographic correlates, magnetic resonance imagine (MRI), and possible mechanics. *Perceptual and Motor Skills,* **94**, 927–949.

29. Carson, S. H., Peterson, J. B., and Higgins, D. M. (2003). Decreased latent inhibition is associated with increased creative achievement in high-functioning individuals. *Personality Processes and Individual Differences,* **85** (3), 499–506; Lubow, R. E. and Gewirtz, J. C. (1995). Latent inhibition in humans: Data, theory, and implications for schizophrenia. *Psychological Bulletin,* **117** (1), 87–103.

Chapter 4: Origins

1. Thomas, K. (1971). *Religion and the Decline of Magic.* London: Weidenfeld & Nicolson.

2. Tolaas, J. and Ullman, M. (1979). Extrasensory communication and dreams. In B. B. Wolman (ed.). *Handbook of Dreams: Research, Theories and Applications.* New York: Van Nostrand Reinhold Company.

3. Tao, W. (1996). Colour terms in Shang oracle bone inscriptions. *Bulletin of the School of Oriental and African Studies,* **59** (1), 63–101.

4. August 2001 issue of *Geology* magazine. URL: news.nationalgeographic.com-news/2001/08/0814_delphioracle.html.

5. Internet Ancient History Sourcebook. URL:http://www.fordham.edu /halsall/ancient/herodotus-creususandsolon.html accessed December 12, 2004.

6. Ebon, M. (1978). History of parapsychology. In M. Ebon (ed.). *The Signet Handbook of Parapsychology.* New York: NAL Penguin, Inc., p. 20.

7. Dodds. F. R. (1971). Supernormal phenomena in classical antiquity. *Proceedings of the Society for Psychical Research,* **55**, 189–237.

8. Ebon, 1978, op. cit., p. 21.

9. Woods, R. K. (1947). *The World of Dreams.* New York: Random House.
10. http://www.law.umkc.edu/faculty/projects/ftrials/ftrials.htm as of January 2005.
11. Gross, C. G. (1997). Emanuel Swedenborg: A neuroscientist before his time. *The Neuroscientist,* **3** (2), 142–147.
12. Lamm, M. (2000). *Emanuel Swedenborg: The Development of His Thought.* West Chester, PA: Swedenborg Foundation Publishers. Also see http://www.swedenborgdigitallibrary.org/ as of January 2005, for many references on Swedenborg's life and thoughts.
13. Alvarado, C. (2001). Concepts of force in early psychical research. In the *Proceedings of Presented Papers: The Parapsychological Association 44th Annual Convention.* (pp. 9–24). Also see: http://www.pesquisapsi.com/content/view/18/41/
14. Crabtree, A. (1993). *From Mesmer to Freud: Magnetic Sleep and the Roots of Psychological Healing.* New Haven, CT: Yale University Press.
15. Crabtree, 1993, op. cit.
16. Beloff, J. (1993). *Parapsychology–A Concise History.* London: Athlone Press, p. 30–31.
17. Ibid, p. 45.
18. Ibid, p. 50.
19. Boring, E. G. (1961). *History, Psychology, and Science: Selected Papers.* New York: John Wiley and Sons, Inc.; Sexton, V. S., and Misiak, H. (1971). *Historical Perspectives in Psychology: Readings.* Belmont, California: Brooks/Cole Publishing Company.
20. Playfair, 2003, op. cit., p. 24.
21. McConnell, R. A. (1976). Parapsychology and physicists. *Journal of Parapsychology,* **40.**
22. Hacking, I. (1988). Telepathy: Origins of Randomization in Experimental Design. *Isis,* **79** (3), p. 438.
23. Warcollier, R. (2001). *Mind to Mind.* Charlottesville, VA: Hampton Roads, p. xii. Originally published in 1948 by Creative Ave. Press.
24. Hacking, 1988, op. cit., p. 440. The word *parapsychology* was coined by German philosopher Max Dessoir, while still a student. See Hovelmann, G. H. (1987). Max Dessoir and the origin of the word "parapsychology," *Journal of the Society for Psychical Research,* **54,** 61–63.
25. Ibid, p. 441
26. See www.cocaine.org, as of March 2005.
27. Van Over, R. (1972). *Psychology and Extrasensory Perception.* New York: New American Library, p. xix.
28. Ehrenwald, J. (1977). Psi, psychotherapy, and psychoanalysis. In B. B.

Wolman (ed.). *Handbook of Parapsychology.* New York: Van Nostrand Reinhold Company, p. 537.

29. Coover also published a report on one of these experiments in 1913. See Coover, J. E. (1913). The feeling of being stared at. *American Journal of Psychology,* **24**, 57–575.

30. Hughes, J. (September 2003). Occultism and the atom: The curious story of isotopes. *Physics World,* 31–35.

31. Mauskopf, S. H., and McVaugh, M. R. (1980) *The Elusive Science: Origins of Experimental Psychical Research.* Baltimore, MD: John Hopkins University Press, p. 56.

32. Hacking, 1988, op. cit., p 450.

33. Jordan, P. (July-August, 1955). Atomic physics and parapsychology. *Newsletter of the Parapsychology Foundation,* **2** (4); Jordan, P. (1951). Reflections on parapsychology, psychoanalysis, and atomic physics, *Journal of Parapsychology,* **4**.

34. Sigmund Freud Museum Vienna, http://www.freud-museum.at/freud/chronolg/1925-e.htm as of January 2005.

35. Tyrrell, G. N. M. (1936–1937). Further research in extra-sensory perception. *Proceedings of the Society for Psychical Research,* **44** (147), 99–166.

36. Wilkins, H. and Sherman, H. M. (2004). *Thoughts through Space.* Charlottesville, VA: Hampton Roads. Originally published in 1951 by C & R Anthony, Inc.

37. Pratt, J. G., Rhine, J. B., Smith, B., Stuart, C., and Greenwood, J. (1940). *Extrasensory Perception after Sixty Years.* Boston: Bruce Humphries Publishers.

38. Carington, W. (1940). Experiments on the paranormal cognition of drawings, *Proceedings of the Society for Psychical Research,* **46**, 34–151.

39. Einstein, A. (1949). Albert Einstein, Philosopher-Scientist. In P. A. Schilpp (ed.), *The Library of Living Philosophers.* Evanston, IL, p. 683.

40. Turing, A. M. (1950). Computing machinery and intelligence. *Mind,* **59**, 433–460; emphasis added.

41. Wolman, B. B. (1977). Mind and body: A contribution to a theory of parapsychological phenomena. In B. B. Wolman (ed.). *Handbook of Parapsychology.* New York: Van Nostrand Reinhold Company, p. 872; Ehrenwald, J. (1977). Psi phenomena and brain research. In B. B. Wolman (ed.). *Handbook of Parapsychology.* New York: Van Nostrand Reinhold Company, p. 709.

42. Stepan, F. (1959). The application of plethysmography to the objective study of so-called extrasensory perception, *Journal of the Society for Psychical Research,* **40**, 162–172.

43. See http://www.foodreference.com/html/html/yearonlytimeline1951–2000.html accessed as of March 17, 2005.

44. Mitchell, E. (1971). An ESP test from Apollo 14. *Journal of Parapsychology*, **35**, 94–111.

45. Child, I. L. (1985). Psychology and anomalous observations: The question of ESP in dreams. *American Psychologist*, **40** (11), 1219–1230.

46. Rabinowitz, D. (May 1990). From the mouths of babes to a jail cell, *Harper's Magazine*.

47. *South China Morning Post*, July 7, 1998, see http://www.forteantimes.com/articles/115_sonypsi.shtml.

48. For a history of Interval, see http://www.wired.com/wired/archive/7.12/interval.html (accessed March 2005).

49. For a history of StarLab, see http://www.space-time.info/starlab/StarlabArchive.html accessed March 2005.

50. Of course, there are many interesting events that can't be repeated upon demand, or aren't stable enough to carefully study. For that portion of the natural world science is ill equipped to pass judgment.

Chapter 5: Putting psi to the test

1. These include Rao, K. R. (2001). *The Basic Experiments in Parapsychology*. (second edition). Jefferson, NC: McFarland & Company; Parker, A. and Brusewitz, G. (2003). A compendium of the evidence for psi. *European Journal of Parapsychology*, **18**, 33–52, and a series called "Studies in Consciousness" by Hampton Roads Press. See www.espresearch.com for a list of these books. They include Upton Sinclair's *Mental Radio*, René Warcollier's *Mind to Mind*, Vasiliev's *Experiments in Mental Suggestion*, and Frederic Myer's *Human Personality and Its Survival of Bodily Death*.

2. The University of Gröningen was founded in 1614; Brugmans, H.I.F.W. (1922). A communication regarding the telepathy experiments in the Psychological Laboratory at Groningen carried out by M. Heymans, Dr. Weinberg, and Dr. H.I.F.W. Brugmans. *Proceedings of the First International Congress of Psychical Research*, Copenhagen, 396–408.

3. Schouten, S. A. and Kelly, E. F. (1978). On the experiments of Brugmans, Heymans, and Weinberg. *European Journal of Parapsychology*, **2**, 247–290.

4. The overall weighted z score for the 188 ESP card studies is $z = 83.1$. This is associated with a probability far less than 10^{-300}. This is based on 2,388,789 hits in 4,599,282 trials, each study adjusted to a 50%-equivalent hit rate; the actual chance hit rates in the card tests ranged from 1% to 50% (100 to 2 symbols per deck). It is very likely that this

database has a selective reporting bias. But it would take 428,000 studies averaging an overall null effect to bring this overall z score down to a nonsignificant level. A more conservative estimate, based on a method recommended by Scargle is 9,800 studies. But that would still take a ratio of 52 unreported, unsuccessful experiments for each known publication to nullify the results. Scargle, J. (2000). Publication bias: The "file-drawer problem" in scientific inference. *Journal of Scientific Exploration*, **14** (1), 91–106.

5. From Table 6.1 in Steinkamp, F. (2005). Forced-choice experiments: Their past and their future. In M. Thalbourne and L. Storm (eds.). *Parapsychology in the Twenty-first Century*. Jefferson, NC: McFarland & Company, Inc., pp. 124–163.

6. I use one standard error bars throughout this book, as they provide an easy way to judge how many standard errors the observed effect is from chance expectation.

7. Rhine, J. B. and Pratt, J. G. (1954). A review of the Pearce-Pratt distance series of ESP tests. *Journal of Parapsychology*, **18**, 165–177.

8. Ibid; Pratt, J. G. and Woodruff, J. L. (1939). Size of stimulus symbols in extrasensory perception. *Journal of Parapsychology*, **3**, 121–158.

9. Hansel, C. E. M. (1966). *ESP: A Scientific Evaluation*. New York: Charles Scribner's Sons.

10. Ibid., p. 127.

11. Ibid., p. 124.

12. Sinclair, U. (2001) *Mental Radio*. Charlottesville, VA: Hampton Roads.

13. Ibid., p. 124.

14. Warcollier, 2001, op. cit.

15. Ibid., p 3.

16. Carington, 1940, op. cit.

17. Ibid., p.129.

18. Ibid., p.131.

19 Tart, C. T., Puthoff, H. E., and Targ, R. (1980). Information transfer under conditions of sensory shielding. *Nature*, **284**, 191; Marks, D. F, and Scott, C. (1986). Remote viewing exposed. *Nature*, **319**, 444.

20. Targ, R. (1994). Remote viewing replication evaluated by concept analysis. *Journal of Parapsychology*, **58**, 271–284.

21. Thousands of research and operational remote viewing sessions were conducted under the auspices of the U.S. government's psi program.

22. Dunne, B. J. and Jahn, R. G. (2003). Information and uncertainty in remote perception research, *Journal of Scientific Exploration*, **17** (2), 207–241.

23. Ibid., p. 229.
24. Bieze, J. (February 1, 2004). In stride with diabetes. *Biomechanics*, p. 19.

Chapter 6: Conscious Psi

1. Shane, S. and Sanger, D. E. (April 1, 2005). Bush panel finds big flaws remain in U.S. spy efforts. *New York Times*.
2. I almost added footnotes to the endnotes, as many levels of devilish details are required to fully appreciate controversial experimental evidence. But while pondering the multiple, hierarchically stacked explanations required to implement such a strategy, I decided that this form of Dante's Inferno would appeal only to Alighieri himself, and so I reluctantly decided to use a single level of endnotes.
3. This maxim is often attributed to astronomer Carl Sagan, but it was first said by the sociologist of science, Marcello Truzzi.
4. Almost 12,000 documents from the formerly classified programs of psi research have been declassified and are available to the public from the Central Intelligence Agency. These programs had exotic code-names such as GRILL FLAME, CENTER LANE, SUN STREAK, and STAR GATE. Ask your friendly neighborhood CIA agent for a set of CDs containing nearly 90,000 pages of reports.
5. This perceptual distortion is similar to the effect of beholding the object of one's romantic love. The physical appearance of a lover is much more attractive to the beholder than to anyone else. Likewise, if the object of one's love is a theory about the way the world works, then evidence that favors the theory will be perceived as attractive.
6. Mullen, B. and Rosenthal, R. (1985). BASIC *Meta-Analysis Procedures and Programs*. Hillsdate, NJ: Lawrence Erlbaum, p. 2.
7. In practice, this relationship isn't so simple. In meta-analyses, study quality tends to increase over time, and later, higher-quality studies also tend to have more samples. Because effect sizes often decrease with higher sample sizes for purely statistical reasons, a negative correlation between study quality and effect size might be due to the co-variation with sample size.
8. A comprehensive meta-analysis involves selecting appropriate effect sizes, examining potential moderating variables, assessing the degree of heterogeneity among the effect sizes, comparing fixed effects vs. random effects models, assessing the impact of selective reporting practices and variations in experimental quality, and so on. While these topics are a delight to statisticians, such details are beyond the scope of this book. Suffice it to say that meta-analysis is an increas-

ingly sophisticated tool used in many scientific disciplines, but like any tool it isn't a panacea.

9. Courtesy of Mrs. Anne Ring, personal correspondence, April 3, 2005.

10. Prasad, J., and Stevenson, I. (1968). A survey of spontaneous psychical experiences in school children of Uttar Pradesh, India. *International Journal of Parapsychology, 10*, 241–261; Rhine, J. B. (1964). *Extra-sensory perception.* Boston, MA: Bruce Humphries.

11. Ullman, M., Krippner, S., and Vaughan, A. (2002). *Dream Telepathy: Experiments in Nocturnal Extrasensory Perception.* Charlottesville, NC: Hampton Roads. Originally published in 1973.

12. The reason a consensus vote is employed to create one trial per night, rather than using the individual ranks of each participant to create multiple trials per night, is to avoid a statistical artifact known as the "stacking effect." This refers to the fact that individual guesses in this sort of experiment might appear to be unrelated, but they're not truly independent. Lack of independence can seriously inflate the statistical evaluation of the data.

13. Sherwood, S. J. and Roe, C. A. (2003). A review of dream ESP studies conducted since the Maimonides dream ESP studies. In J. Alcock, J. Burns and A. Freeman (eds.). *Psi Wars: Getting to Grips with the Paranormal.* Thorverton, UK: Imprint Academic.

14. Child, 1985, op. cit., and Ullman, Krippner and Vaughan, 2002, op. cit.

15. The methods used here are standard meta-analytic techniques employed throughout the behavior, social, and medical sciences.

16. More precisely, Rosenthal's method calculates the number of studies required to bring the overall level of statistical significance to $p > 0.05$.

17. The reason is that by convention, "statistically significant" means that we've obtained a positive outcome with odds against chance greater than 20 to 1. This is equivalent to a probability less than 5% (usually denoted as $p < 0.05$). But there's another way to get a statistically significant result, namely with a strong *negative* outcome. Say we ran a dream psi study and ended up with a 0% hit rate. Such a low hit rate is most unlikely to occur by chance, and so we might be motivated to publish that study even though it resulted in an outcome strongly in the "wrong" direction. But since most studies are not predicted in advance to end up with a strong negative result, the probability of that result has to be adjusted. The usual adjustment for results predicted without regard to direction is a factor of 2. Thus, we can imagine that studies with a strong *positive* outcome associated with $p < 0.05$, and

also studies with a strong *negative* outcome associated with $p < 0.025$, would all be published. This asymmetry creates an overall estimated distribution with an overall slightly negative average, rather than a zero average. See Hsu, L. M. (2002). Fail-safe Ns for one- versus two-tailed tests lead to different conclusions about publication bias. *Understanding statistics,* 1 (2), 85–100. Also see Scargle, 2000, op. cit.

18. Using Hsu's model (ibid.) the file drawer ranges from 1,339 to 2,697, or ratios of 28:1 to 57:1.

19. There are more formal ways of making this assessment. E.g., we can see if a funnel plot is lopsided by calculating the correlation between sample size and effect size. A lopsided funnel will show up as a significant negative correlation. The resulting correlation, shown as the dotted line, is basically flat: $r = -0.014$, $p = 0.46$.

20. Throughout this book effect size refers to $e = z / \sqrt{N}$, where N is the number of trials in an experiment, and z is a standard normal deviate distributed as N[0,1]. The experimental results discussed in this chapter are actually based on binomial distributions that are closely approximated by the normal distribution because N is large. Thus the z score is used as an estimate of the exact binomial probabilities, and likewise the effect size e is an estimate. The overall mean effect size is $e = 0.18 \pm 0.03$ (weighted average plus or minus the standard error) where $e = 0$ is expected by chance. The mean is weighted by sample size, based on a fixed effects model, which assumes that the underlying effects in the different studies are approximately the same. See Hedges, L. V. (1994). Fixed effect models. In L. V. Hedges and H. Cooper (eds.), *The Handbook of Research Synthesis* (pp. 285–299). New York: Russell Sage Foundation; also Hedges, L. V., and Vevea, J. L. (1998). Fixed- and random-effects models in meta-analysis. *Psychological Methods,* 3, 486–504.

21. Bertini, M., Lewis, H., and Witkin, H. (1964). Some preliminary observations with an experimental procedure for the study of hypnagogic and related phenomena. *Archivo di Psicologia Neurologia e Psychiatra,* 6, 493–534.

22. Palmer, J. (2003). ESP In the Ganzfeld: Analysis of a debate. In Alcock, J., Burns, J., and Freeman, A. (eds.). *Psi Wars: Getting to Grips with the Paranormal.* Thorverton, UK: Imprint Academic.

23. Bem, D. J., and Honorton, C. (1994). Does psi exist? Replicable evidence for an anomalous process of information transfer. *Psychological Bulletin,* 115, 4–18; Bem is a prominent social psychologist and a member of the Psychic Entertainers Association, an organization of entertainers who specialize in simulating psychic ability. Honorton was a

brilliant psi researcher working on his doctorate at the University of Edinburgh in 1992 when he tragically died of a heart attack.

24. Milton, J. and Wiseman, R. (1999). Does psi exist? Lack of replication of an anomalous process of information transfer. *Psychological Bulletin,* **125**, 387–391.

25. Had they simply added up the total number of hits and trials conducted in those 30 studies, they would have found a statistically significant result with one-tailed odds against chance of about 20 to 1.

26. Storm, L. and Ertel, S. (2001). Does psi exist? Milton and Wiseman's (1999) meta-analysis of ganzfeld research. *Psychological Bulletin,* **127**, 424–433.

27. Milton, J., and Wiseman, R. (2001). Does psi exist? Reply to Storm and Ertel (2001). *Psychological Bulletin,* **127**, 434–438.

28. Milton, J. (1999). Should ganzfeld research continue to be crucial in the search for a replicable psi effect? Part I. Discussion paper and an introduction to an electronic mail discussion. *Journal of Parapsychology,* **63**, 309–333.

29. Dalton, K. (1997). Exploring the links: Creativity and psi in the ganzfeld. *Proceedings of Presented Papers: The Parapsychological Association 40th Annual Convention,* pp. 119–134.

30. Bem, D. J., Palmer, J., and Broughton, R. S. (2001). Updating the ganzfeld database: A victim of its own success? *Journal of Parapsychology,* **65**, 207–218.

31. To ensure there was no bias in this analysis, they asked independent judges, who were blind to the outcomes of the studies, to perform the assessment of "standardness."

32. The debate over potential flaws in these experiments has been extremely thorough. Examples of suggested flaws include (a) whether any form of "sensory leakage" could have passed between the senders, receivers, or experimenters, (b) whether fingerprints on paper targets due to handling by the sender might have provided a clue to the receiver about which target was the real one, (c) whether having only one experimenter might represent a security flaw, (d) whether the targets were adequately randomized across separate trials, and (e) whether the order in which the target picture was observed by the receiver during the judging process was adequately randomized. John Palmer concluded his exhaustive review of the ganzfeld debate as follows: "My impression over many years is that critics of parapsychology are very good at providing *conceivable* normal explanations for psi effects but rather poor at providing *plausible* normal explanations for them."

33. This excludes a few of the earliest ganzfeld studies that couldn't be evaluated with a hit vs. miss type of analysis.

34. The overall weighted effect size is $e = 0.16 \pm 0.02$.

35. This uses the file-drawer method recommended in Scargle (2000), op. cit.

36. Jennions, M. D. and Müller, A. P. (2002). Relationships fade with time: a meta-analysis of temporal trends in publication in ecology and evolution. *Proceedings of the Royal Society, Biological Sciences,* **269,** 43–48.

37. Sanchez, J. Dohoo, I., Carrier, J., and DesCôteaux, L. (2004). A meta-analysis of the milk-production response after anthelmintic treatment in naturally infected adult dairy cows. *Preventive Veterinary Medicine.*

38. S. Eidelman et al., (2004). Review of Particle Physics. *Physics Letters B,* **592,** 1.

39. Google search on November 27, 2004.

40. In more sophisticated designs, Jack consults a prepared table of random assignments or uses a truly random number generator.

41. Sheldrake, R. (1998). The sense of being stared at: Experiments in schools. *Journal of the Society of Psychical Research,* **62,** 311–323; Sheldrake, R. (1999). The "sense of being stared at" confirmed by simple experiments. *Biology Forum,* **92,** 53–76; Sheldrake, R. (2000). The "sense of being stared at" does not depend on known sensory clues. *Biology Forum,* **93,** 209–224; Sheldrake, R. (2001). Experiments on the sense of being stared at: The elimination of possible artefacts. *Journal of the Society for Psychical Research,* **65,** 122–137; Sheldrake, R. (2003). *The Sense of Being Stared At.* New York: Crown Publishers.

42. Coover, J. E. (1913). The feeling of being stared at. *American Journal of Psychology,* **24,** 57–575; Poortman, J. J. (1959). The feeling of being stared at. *Journal of the Society for Psychical Research,* **40,** 4–12; Radin, D. I. (2004). On the sense of being stared at: An analysis and pilot replication. *Journal of the Society for Psychical Research,* **68,** 246–253; Nelson, L. A. and Schwartz, G. E. (2005). Human biofield and intention detection: Individual differences. *Journal of Alternative and Complementary Medicine,* **11** (1), 93–101.

43. Colwell, J., Schröder, S., and Sladen, D. (2000). The ability to detect unseen staring: A literature review and empirical tests. *British Journal of Psychology,* **91,** 71–85; Marks, D. and Colwell, J. (2000). The psychic staring effect: An artifact of pseudo randomization. *Skeptical Inquirer,* September/October, 41–49.

44. Duval, S. J., and Tweedie, R. L. (2000). A nonparametric "trim and fill" method of accounting for publication bias in meta-analysis. *Journal of the American Statistical Association,* **95,** 89–98.

45. A more comprehensive meta-analysis of these studies shows that effects sizes for the subset of higher security, through-the-window experiments, without trial-by-trial feedback, is homogenous. Under a fixed effects model the weighted effect size is $e = 0.060 \pm 0.007$, $p = 4.8 \pm 10^{-17}$. See Radin, D. I. (in press, 2005). The sense of being stared at: A preliminary meta-analysis. *Journal of Consciousness Studies.*

46. The original effect size of $e = 0.089 \pm 0.005$ with 33,357 known trials is adjusted to $e = 0.078 \pm 0.005$ with 34,097 estimated trials.

47. Besides implicit learning of sensory cues, another explanation for the better outcomes of studies conducted with Jack and Jill in close proximity include bioelectromagnetic or "biofield" interactions.

48. Nelson and Schwartz, 2005, op. cit.

Chapter 7: Unconscious Psi

1. Some of these chambers are also magnetically shielded.

2. More commonly the measurement is skin conductance, the inverse of resistance. Skin conductance reflects the activity of the autonomic nervous system.

3. In similar studies other physiological parameters are also measured.

4. For example, the calm-activate order must be randomly counterbalanced in such a way as to prevent natural drifts in Jill's physiology from accidentally matching that order, the investigator should be blind to this order, and sound, vibration, and electromagnetic checks should be run to confirm that no ordinary information can reach Jill.

5. Braud, W. G., Shafer, D., and Andrews, S. (1993). Further studies of autonomic detection of remote staring, new control procedures, and personality correlates. *Journal of Parapsychology,* **57,** 391–409; Braud, W. G., Shafer, D., and Andrews, S. (1993). Reactions to an unseen gaze (remote attention): A review, with new data on autonomic staring detection. *Journal of Parapsychology,* **57,** 373–390.

6. Bootstrap statistics, or other nonparametric permutation methods, are typically used to assess the statistical significance of these differences.

7. Schmidt, S., Schneider, R., Utts, J., and Walach, H. (2004). Distant intentionality and the feeling of being stared at: Two meta-analyses. *British Journal of Psychology,* **95,** 235–247.

8. Cohen's *d* weighted effect size $d = 0.11$, $p = 0.001$.

9. Cohen's $d = 0.13$, $p = 0.01$.

10. Schmidt, Schneider, Utts and Walach, 2004, op. cit.

11. Duane, T. D. and Behrendt, T. (1965). Extrasensory electroencephalographic induction between identical twins. *Science,* **150**: 367; Tart, C. T. (1963). Possible physiological correlates of psi cognition. *International Journal of Parapsychology,* **5**, 375–386.

12. Hearne, K. (1977). Visually evoked responses and ESP. *Journal of the Society for Psychical Research,* **49**, 648–657; Hearne, K. (1981). Visually evoked responses and ESP: Failure to replicate previous findings. *Journal of the Society for Psychical Research,* **51**, 145–147; Kelly, E. F. and Lenz, J. (1976). EEG changes correlated with a remote stroboscopic stimulus: A preliminary study. In Morris, J., Roll, W., and Morris, R. (eds.). *Research in Parapsychology 1975,* Metuchen, NJ: Scarecrow Press, p. 58–63 (abstracted in *Journal of Parapsychology,* 1975, **39**, 25); Lloyd, D. H. (a pseudonym) (1973). Objective events in the brain correlating with psychic phenomena. *New Horizons,* 1, 69–75; May, E. C., Targ, R. and Puthoff, H. E. (2002). EEG correlates to remote light flashes under conditions of sensory shielding. In Tart, C. T., Puthoff, H. E., and Targ, R. (eds.). *Mind at large: IEEE symposia on the nature of extrasensory perception.* Charlottesville, VA: Hampton Roads Publishing Company, 1979/2002; Millar, B. (1975). An attempted validation of the "Lloyd effect." In Morris, J. D., Roll, W. G., and Morris, R. L. (eds.). *Research in Parapsychology 1975,* Metuchen, NJ: Scarecrow Press, 25–27; Millay, J. (1999). *Multidimensional Mind: Remote Viewing in Hyperspace.* Berkeley, CA: North Atlantic Books; Orme-Johnson, D. W., Dillbeck, M. C., Wallace, R. K. and Landrith, G. S. (1982). Intersubject EEG coherence: Is consciousness a field? *International Journal of Neuroscience,* **16**, 203–209; Rebert, C. S. and Turner, A. (1974). EEG spectrum analysis techniques applied to the problem of psi phenomena. *Behavioral Neuropsychiatry,* **6**, 18–24; Targ, R. and Puthoff, H. (1974). Information transmission under conditions of sensory shielding. *Nature,* **252**, 602–607.

13. Grinberg-Zylberbaum, J., Delaflor, M., Attie, L., and Goswami, L. (1994). The Einstein-Podolsky-Rosen paradox in the brain: The transferred potential. *Physics Essays,* **7**, 422–428; Grinberg-Zylberbaum, J., Delaflor, M., Sanchez, M. E. and Guevara, M. A. (1993). Human communication and the electrophysiological activity of the brain. *Subtle Energies and Energy Medicine,* **3**, 25–43; Grinberg-Zylberbaum, J. and Ramos, J. (1987). Patterns of interhemispheric correlation during human communication. *International Journal of Neuroscience,* **36**, 41–53.

14. Sabell, A., Clarke, C., and Fenwick, P. (2001). Inter-Subject EEG correlations at a distance–the transferred potential. In: Alvarado, C. S., ed. *Proceedings of the 44th Annual Convention of the Parapsychological Association,* New York, NY, pp. 419–422; Fenwick, B. C. P., Vigus, N., and Sanders, S. (1998). The transferred potential (unpublished manuscript); Standish, L. J., Johnson, L. C., Richards, T. and Kozak, L. (2003). Evidence of correlated functional MRI signals between distant human brains. *Alternative Therapies in Health and Medicine,* 9, 122–128; Standish, L. J., Kozak, L., Johnson, L. C., and Richards, T. (2004). Electroencephaolographic evidence of correlated event-related signals between the brains of spatially and sensory isolated human subjects. *Journal of Alternative and Complementary Medicine,* 10, 307–314; Wackermann, J., Seiter, C., Keibel, H., and Walach, H. (2003). Correlations between brain electrical activities of two spatially separated human subjects. *Neuroscience Letters,* 336, 60–64.

15. Wackermann et al., 2003, op. cit.

16. The replication was led by Todd Richards, a professor of radiology at the University of Washington. The article reporting this replication was under peer review for publication as this book went to press.

17. This study was actually performed before Standish's fMRI experiment, but was published afterwards.

18. Kittenis, M., Caryl, P., and Stevens, P. (2004). Distant psychophysiological interaction effects between related and unrelated participants, *Proceedings of the Parapsychological Association Convention,* pp. 67–76.

19. Radin, D. I. (2004). Event-related EEG correlations between isolated human subjects. *Journal of Alternative and Complementary Medicine,* 10, 315–324. My thanks to laboratory assistants Jenny Matthews, Charlene Farrell and Gail Hayssen.

20. The shielded room was a Series 81 Solid Cell chamber, made by Lindgren/ETS of Cedar Park, Texas.

21. We looked at Jack and Jill's EEG record from the moments of video onset and offset plus and minus five seconds. Movement and eyeblink artifacts in the EEG record were removed and then the ensemble variance across the EEG records were determined across all participants.

22. The correlation was $r = 0.20$, $p = 0.0002$; the probability was determined using bootstrap nonparametric analysis.

23. The control correlation was $r = -0.03$, $p = 0.61$, one-tail.

24. These curves represent a measure of EEG variance, averaged across all participants and all stimuli.

Chapter 8: Gut feelings

1. It was clear from the context of the survey that this meant intuitive hunches, and not simply meaningless somatic sensations.

2. Torff, B. and Sternberg, R. J. (2001). Intuitive conceptions among learners and teachers. In Torff, B., and Sternberg, R. J. (eds.). *Understanding and Teaching the Intuitive Mind: Student and Teacher Learning.* Mahwah, NJ: Lawrence Erlbaum Associates, Publishers, 3–26; Damasio, A. R. (1994). *Descartes' Error: Emotion, Reason, and the Human Brain.* New York: G. P. Putnam's Sons; Damasio, A. R. (1996). The somatic marker hypothesis and the possible functions of the prefrontal cortex. *Philosophical Transactions of the Royal Society of London,* B, 351, 1413,1420.

3. Hams, S. P. (2000). A gut feeling? Intuition and critical care nursing. *Intensive Critical Care Nursing,* 16, 310–318.

4. Stern, R. M., Ray, W. J., and Quigley, K. S. (2001). *Psychophysiological Recording,* 2nd ed. New York: Oxford University Press; Stern, R. M. (1985). A brief history of the electrogastrogram. In Stern, R. M., and Koch, K. L. (eds.). *The Electrogastrogram: Research studies and applications.* New York: Praeger, pp. 3–9.

5. Muth, E. R., Koch, K. L., Stern, R. M., and Thayer, J. F. (1999). Effect of autonomic nervous system manipulations on gastric myoelectrical activity and emotional responses in healthy human subjects. *Psychosomatic Medicine,* 61, 297–303.

6. Beaumont, W. (1833). *Experiments and Observations on the Gastric Juice and Physiology of Digestion.* Plattsburg, NY: Allen, F. P., Sadler, H. H., and Orten, A. U. (1968). The complementary relationship between the emotional state and the function of the ileum in a human subject. *American Journal of Psychiatry,* 124, 1375–1384; Katkin, E. S., Wiens, S., and Ohman, A. (2001). Nonconscious fear conditioning, visceral perception, and the development of gut feelings. *Psychological Science,* 12, 366–370; Houghton, L. A., Calvert, E. L., Jackson, N. A., Cooper, P., and Whorwell, P. J. (2002).Visceral sensation and emotion: a study using hypnosis. *Gut,* 51, 701–704; Mayer, E. A., Naliboff, B., and Munakata, J. (2000). The evolving neurobiology of gut feelings. *Progress in Brain Research,* 122, 195–206; Welgan, P., Meshkinpour H, and Beeler, M. (1988). Effect of anger on colon motor and myoelectric activity in irritable bowel syndrome. *Gastroenterology,* 94, 1150–1156.

7. Radin, D. I., and Schlitz, M. J. (2005). Gut feelings, intuition, and emotions: An exploratory study. *Journal of Alternative and Complementary Medicine,* in press.

Chapter 9: Mind-Matter Interaction

1. Siegfried, T. (2000). *The Bit and the Pendulum: From Quantum Computing to M Theory–The New Physics of Information*. New York: John Wiley & Sons; Fredkin, E. (2003). An introduction to digital philosophy. *International Journal of Theoretical Physics*, **42** (2), 189–247. Also see http://digitalphysics.org/.

2. Wheeler, J. A. (1990). Information, physics, quantum: the search for links. In: Zurek W. H., ed. *Complexity, Entropy, and the Physics of Information*. Santa Fe Institute Studies in the Sciences of Complexity, vol. VIII. Reading, Massachusetts: Perseus Books.

3. Rhine, J. B. (1944). "Mind over matter" or the PK effect. *Journal of the American Society for Psychical Research*, **38**, 185–201.

4. Girden, E., Murphy, G., Beloff, J., Flew, A., Rush, J. H., Schmeidler, G., and Thouless, R. H. (1964). A discussion of "A review of psychokinesis (PK)." *International Journal of Parapsychology*, **6**, 26–137; Girden, E., and Girden, E. (1985). Psychokinesis: Fifty years afterward. In P. Kurtz (ed.), *A Skeptic's Handbook of Parapsychology*. Buffalo, NY: Prometheus Press, 129–146; Girden, E. (1962). A review of psychokinesis (PK). *Psychological Bulletin*, **59**, 353–388; Girden, E., Murphy, G., Beloff, J., Flew, A., Rush, J. H., Schmeidler, G., and Thouless, R. H. (1964). A discussion of "A review of psychokinesis (PK)." *International Journal of Parapsychology*, **6**, 26–137.

5. Radin, D. I., and Ferrari, D. C. (1991). Effects of consciousness on the fall of dice: A meta-analysis. *Journal of Scientific Exploration*, **5**, 61–84.

6. Weighted mean effect size $e = 0.0122 \pm 0.0006$.

7. Adjusted weighted mean effect size $e = 0.0114 \pm 0.0006$, with an estimated 2.64 million dice thrown.

8. This is supported by a study by biologist Rupert Sheldrake, who found in a large survey of experiments published in the physical, biological, medical, and psychological sciences that standard protections for experimenter bias, the double-blind design, were very rarely used in mainstream disciplines but used frequently in parapsychological studies. See Sheldrake, R. (1998). Experimenter effects in scientific research: How widely are they neglected? *Journal of Scientific Exploration*, **12** (1), 73–78.

9. Multiple $R = 0.25$, $F(10,137) = 0.95$, $p = 0.49$.

10. Of course, I don't mean literally changing the shape of the die, but rather altering the *a priori* probabilities associated with each die face.

11. This is due to the relationship $z = e\sqrt{N}$, where z tells us how far we are from chance expectation, e is the PK effect size per die, and N is the number of dice thrown at once.

12. $r = 0.69$, $p = 0.009$ one-tail, of the fit between the expected and observed curves.

13. $r = 0.92$, $p = 0.0002$.

14. This fit is a second-order polynomial, $y = 0.0007x^2 + 0.0172x - 0.0132$.

15. Jahn, R. G., Dunne, B. J., Nelson, R. D., Dobyns, Y. H., and Bradish, G. J. (1997). Correlations of random binary sequences with pre-stated operator intention: A review of a 12-year program. *Journal of Scientific Exploration,* **11** (3), 345–367.

16. The actual figure is $p = 3.5$ x 10^{-13}, more than seven standard errors from chance.

17. Jahn, R., Dunne, G., Bradish, Y., Dobyns, A., Lettieri, A., Nelson, R., Mischo, J., Boller, E., Bösch, H., Vaitl, D., Houtkooper, J., and Walter, B. (2000). Mind/machine interaction consortium: PortREG replication experiments. *Journal of Scientific Exploration,* **14** (4), 499–555.

18. Correlation between earlier PEAR results and portREG $r = 0.988$, $p = 0.048$. Of course, when correlating only three points we must be very careful with interpretations, nevertheless the relationship is intriguing.

19. The original PEAR results were collapsed into two datapoints, one each for the HI and LO aims, and the mega-trial results were collapsed into 6 datapoints corresponding to the HI and LO aims for each of the three replication sites: PEAR, Freiburg and Giessen.

20. The overall weighted effect size is $e = 0.000122 \pm 0.000028$.

21. In the adjusted funnel plot, the weighted effect size is $e = 0.000101 \pm 0.000030$.

22. Following the method of Hsu, we find the answer ranges between 2,507 to 4,396, or a ratio from 5:1 to 9:1. We also found that there was no significant relationship between study quality and effect size, $r = -0.06$, $t = -1.23$, $N = 490$, $p = 0.11$ (one tail).

23. I am indebted to Princeton University physicist York Dobyns for assistance in gathering these figures; Dobyns, Y. H. (1996). Selection versus influence revisited: New methods and conclusions. *Journal of Scientific Exploration,* **10** (2), 253–268; Dobyns, Y. H. (2000). Overview of several theoretical models on PEAR data. *Journal of Scientific Exploration,* **14** (2), 163–194; Dobyns, Y. H. and Nelson, R. D. (1998). Empirical evidence against Decision Augmentation Theory. *Journal of Scientific Exploration,* **12** (2), 231–257.

24. May, E. C., Utts, J. M., and Spottiswoode, S. J. P. (1995). Decision augmentation theory: applications to the random number generator database. *Journal of Scientific Exploration,* **9**, 4, 453.

25. Ibison, M. (1998). Evidence that anomalous statistical influence de-

pends on the details of the random process. *Journal of Scientific Exploration*, **12**, 407–423.

26. My guess is that this is what's driving the correlations observed between brain activity and consciousness.

Chapter 10: Presentiment

1. Honorton, C. and Ferrari, D. C. (1989). Future telling: A meta-analysis of forced-choice precognition experiments, 1935–1987, *Journal of Parapsychology*, **53**, 281–308.

2. This is based on Rosenthal's method, not the more conservative Scargle or Hsu methods used elsewhere in this book.

3. Steinkamp, F., Milton, J., and Morris, R. L. (1998). Meta-analysis of forced-choice experiments comparing clairvoyance and precognition. *Journal of Parapsychology*, **62**, 193–218.

4. These odds are based on quality-weighted Stouffer Z scores, ibid., Table 3, p. 202. The large difference in odds is largely due to the fact that the precognition studies had more statistical power.

5. Good, I. J. (1961). Letter to the editor. *Journal of Parapsychology*, **25**, p. 58.

6. Levin, J. and Kennedy, J. (1975). The relationship of slow cortical potentials to psi information in man. *Journal of Parapsychology*, **39**, 25–26.

7. Hartwell, J. W. (1978). Contingent negative variation as an index of precognitive information. *European Journal of Parapsychology*, **2**, 83–103.

8. Hartwell, J. W. (1979). An extension to the CNV study and an evaluation. *European Journal of Parapsychology*, **2** (4), 358–364.

9. Vassy, Z. (1978). Method for measuring the probability of one bit extrasensory information transfer between living organisms. *Journal of Parapsychology*, **42**, 158–160; Vassy, Z. (in press). A study of telepathy by classical conditioning, *Journal of Parapsychology*.

10. Radin, D. I. (2004). Electrodermal presentiments of future emotions. *Journal of Scientific Exploration*, **18**, 253–274.

11. Average normalized change in SCL (ΔSCL) across all trials, which are split into two equal datasets, "calm" and "emotional" according to each trial's pre-assessed target emotionality ratings. Nonparametric randomized permutation analysis indicates that the difference in pre-stimulus curves is associated with $z = 2.92$, $p = 0.002$ (one-tailed).

12. Ito, T.A., Cacioppo, J. T., and Lang, P. J. (1998). Eliciting affect using the International Affective Picture System: Bivariate evaluation and ambivalence. *Personality and Social Psychology Bulletin*, **24**, 856–879; Lang, P. J., Bradley, M. M., and Cuthbert, B. N. (1995). *International Affective*

Picture System (IAPS): Technical manual and affective ratings. Gainesville, FL: Center for Research in Psychophysiology, University of Florida.

13. Radin 2004, op. cit.; Radin, D. I. (1997). Unconscious perception of future emotions: An experiment in presentiment. *Journal of Scientific Exploration,* **11** (2), 163–180.

14. To be clear, this is not a correlation between prestimulus vs. poststimulus skin conductance, as the autocorrelations in those figures would guarantee a large positive correlation. Instead, this is a correlation between prestimulus skin conductance vs. pre-assessed emotionality ratings for each image. $r = 0.04$, $t = 2.42$, $N = 4{,}569$, $p = 0.008$.

15. For example: Bierman, D. J. and Radin, D. I. (1997). Anomalous anticipatory response on randomized future conditions. *Perceptual and Motor Skills,* **84**, 689–690; Bierman, D. J., and Radin, D. I. (1998). Conscious and anomalous nonconscious emotional processes: A reversal of the arrow of time? *Toward a Science of Consciousness, Tucson III.* MIT Press, 1999, 367–386; Bierman, D. J., and Scholte, H. S. (2002). Anomalous anticipatory brain activation preceding exposure of emotional and neutral pictures. Paper presented at Toward a Science of Consciousness, Tucson IV; McCraty, R., Atkinson, M., and Bradley, R.T. (2004). Electrophysiological evidence of intuition: Part 1. The Surprising role of the heart. *Journal of Alternative and Complementary Medicine,* **10**, 133–143; McCraty, R., Atkinson, M., and Bradley, R.T. (2004). Electrophysiological evidence of intuition: Part 2. A system-wide process? *Journal of Alternative and Complementary Medicine,* **10**, 325–336; Norfolk, C. (1999). Can future emotions be perceived unconsciously? An investigation into the presentiment effect with reference to extraversion. Unpublished manuscript, Department of Psychology, University of Edinburgh; Spottiswoode, S. J. P., and May, E. C. (2003). Skin conductance prestimulus response: analyses, artifacts and a pilot study. *Journal of Scientific Exploration,* **17** (4), 617–641; Wildey, C. (2001). *Impulse Response of Biological Systems.* Master's Thesis, Department of Electrical Engineering, University of Texas at Arlington; Parkhomtchouk, D. V., Kotake, J., Zhang, T., Chen, W., Kokubo, H., and Yamamoto, M. (2002). An attempt to reproduce the presentiment EDA response. *Journal of International Society of Life Information Science,* **20** (1), 190–194.

16. Reported as $p = 0.13$ for humans, $p = 0.15$ for worms. I transformed these p values into one-tailed z scores and combined them as Stouffer $z = 1.53$, $p = 0.06$.

17. McCraty et al., 2004, op. cit.

18. From McCraty, Part I, 2004, op. cit.

19. Broughton, R. S., Kanthamani, H., and Khilji, A. (1989). Assessing the PRL success model on an independent ganzfeld data base. Paper presented at the Parapsychological Association 32nd Annual Convention, San Diego, California; Honorton, C., Berger, R. E., Varvoglis, M. P., Quant, M., Derr, P., Schecter, E. I., and Ferrari, D. C. (1990). Psi communication in the ganzfeld: Experiments with an automated testing system and a comparison with a meta-analysis of earlier studies. *Journal of Parapsychology,* **54** (2), 99–139; Honorton, C. (1992). The ganzfeld novice: Four predictors of initial ESP performance. Paper presented at the Parapsychological Association 35th Annual Convention, Las Vegas, NV.

20. Broughton, R. S., and Alexander, C. H. (1996). Autoganzfeld II: An attempted replication of the PRL ganzfeld research. Paper presented at the Parapsychological Association 39th Annual Convention, San Diego, CA.

21. They analyzed the data in two ways. One method used the skin conductance averaging technique employed in most of the previous studies; the second method compared the number of skin conductance responses occurring before the audio stimuli vs. before the silent stimuli. This response is observed as a small fluctuation in the skin conductance record (technically it's called a "non-specific skin conductance response" or NS-SCR).

22. Personal communication, Vassy, Z., December 14, 2004.

23. Bierman, D. J. and Radin, D. I. (1997). Anomalous anticipatory response on randomized future conditions. *Perceptual and Motor Skills,* **84**, 689–690; Bierman, D. J. and Radin, D. I. (1998). Conscious and anomalous nonconscious emotional processes: A reversal of the arrow of time? *Toward a Science of Consciousness, Tucson III.* MIT Press, 1999, 367–386; Bierman, D. J. (2001). New developments in presentiment research, or the nature of time. Presentation to the Bial Foundation Symposium, Porto, Portugal.

24. Bechara, A., Damasio, A. R., Damasio, H., and Anderson, S. W. (1994). Insensitivity to future consequences following damage to human prefrontal cortex. *Cognition,* **50**, 7–15; Bechara, A., Tranel, D., Damasio, H., and Damasio, A. R. (1996). Failure to respond autonomically to anticipated future outcomes following damage to prefrontal cortex. *Cerebral Cortex,* **6**, 215–225; Globisch, J., Hamm, A.O., Estevez, F., and Öhman, A. (1999). Fear appears fast: Temporal course of startle reflex potentiation in animal fear subjects. *Psychophysiology,* **36**, pp. 66–75.

25. Bierman, D. J. (2000). Anomalous baseline effects in mainstream emotion research using psychophysiological variables. *Proceedings of Presented Papers: The 43rd Annual Convention of the Parapsychological Association,* 34–47.

26. Bierman and Scholte, 2002, op. cit.

27. T-test of the difference between erotic vs. neutral = 2.89, $df = 39$, $p<0.01$ three seconds before the stimulus.

28. Bierman and Scholte, 2002, op. cit.

Chapter 11: Gaia's Dreams

1. Teilhard de Chardin, P. (1975). *The Phenomenon of Man.* New York: Harper & Row. Originally published in French in 1955 by Editions du Seuil, Paris.

2. Gaia was a fictional planet in Isaac Asimov's 10-volume science fiction epic known as the Foundation Series. Asimov's Gaia planet had a global consciousness, with all beings and even inanimate objects connected via telepathy. Gaia had aspirations of evolving into a galactic consciousness.

3. Nelson, R. D., Bradish, G. J., Dobyns, Y. H., Dunne, B. J., and Jahn, R. G. (1996). FieldREG anomalies in group situations. *Journal of Scientific Exploration,* 10, 111–142; Nelson, R. D., Jahn, R. G., Dunne, B. J., Dobyns, Y. H., and Bradish, G. J. (1998). FieldREG II: Consciousness field effects: Replications and explorations. *Journal of Scientific Exploration,* 12, 425–454; Radin, D. I. (1997). *The Conscious Universe.* San Francisco: HarperEdge; Blasband, R. A. (2000). The ordering of random events by emotional expression. *Journal of Scientific Exploration,* 14, 195–216; Radin, D. I., Rebman, J. M., and Cross, M. P. (1996). Anomalous organization of random events by group consciousness: Two exploratory experiments. *Journal of Scientific Exploration,* 10, 143–168; Yoichi, H., Kokubo, H. and Yamamoto, M. (2002). Anomaly of random number generator outputs: Cumulative deviation at a meeting and New Year's holiday. *Journal of International Society of Life Information Science,* 20 (1), 195–201; Yoichi, H., Kokubo, H., and Yamamoto, M. (2004). Anomaly of random number generator outputs (II): Cumulative deviation at New Year's holiday. *Journal of International Society of Life Information Science,* 22 (1), 142–146; Bierman, D. J. (1996). Exploring correlations between local emotional and global emotional events and the behavior of a random number generator. *Journal of Scientific Exploration,* 10, 363–373; Kokubo, H., Yoichi, H., and Yamamoto, M. (2002). Data analyses of a field number generator.

The Japanese Journal of Parapsychology, **7**, 11–16. (In Japanese); Hirukawa, T., and Ishikawa, M. (2004). Anomalous fluctuation of RNG data in Nebuta: Summer festival in Northeast Japan. *Proceedings of Presented Papers,* Parapsychological Association 2004; Hagel, J., and Tschapke, M. (2004). The local event detector (LED): An experimental setup for an exploratory study of correlations between collective emotional events and random number sequences, *Proceedings of Presented Papers,* Parapsychological Association 2004.

4. Rowe, W. D. (1998). Physical measurements of episodes of focused group energy. *Journal of Scientific Exploration,* **12**, 569–583.

5. Ibid.

6. Nelson et al., 1996, op. cit.; Nelson et al., 1998, op. cit.

7. Schlitz, M., Radin, D. I., Malle, B. F., Schmidt, S., Utts, J., and Yount G. L. (2003). Distant healing intention: Definitions and evolving guidelines for laboratory studies. *Alternative Therapies in Health and Medicine,* **9**, A31–A43.

8. Crawford, C. C., Jonas, W. B., Nelson, R., Wirkus, M., and Wirkus, M. (2003). Alterations in random event measures associated with a healing practice. *Journal of Alternative and Complementary Medicine,* **9** (3), 345–353; Schwartz, G. E. R., Russek, L. G. S., Zhen-Su, S., Song, L.Z.Y.X., and Xin, Y. (1997). Anomalous organization of random events during an international quigong meeting: Evidence for group consciousness or accumulated qi fields? *Subtle Energies & Energy Medicine,* **8**, 55–65.

9. Most experiments in the behavioral, social, and medical sciences demonstrate correlations and not causes.

10. This isn't quite true, because a living and a nonliving system can be thought of as two separate RNGs, and just because two RNGs simultaneously respond to distant intention that doesn't definitively prove a causal relationship. However, as the number of independent physical systems responding to intention increases, the evidence for a causal influence also increases.

11. Ansaloni, A. (2003). Effect of Lourdes water on water pH. *Bollettino chimico farmaceutico,* 2003, 142:202–205; Ansaloni, A. (2002). Effect of the Lourdes water on "chlorinated water." *Bollettino chimico farmaceutico,* **141**, 80–83; Cohen, K. (2003). Where healing dwells: The importance of sacred space. *Alternative Therapies in Health and Medicine,* **9**, 68–72; Devereux, P. (1999). *Places of Power: Measuring the Secret Energy of Ancient Sites,* 2nd edition. London: Blandford Press; Szabo, J. (2002). Seeing is believing? The form and substance of French medical debates over Lourdes. *Bulletin of the History of Medicine,* **76**, 199–230.

12. Roll W G, and Persinger M. A. (2001). Investigations of poltergeists and haunts: A review and interpretation. In J. Houran, and R. Lange, (eds.). *Hauntings and Poltergeists: Multidisciplinary Perspectives.* Jefferson, NC: McFarland, pp. 123–163.

13. See http://www.randomnumbergenerator.nl/rng/home.html (accessed February, 2005) for the ORION RNG, and http://noosphere.princeton .edu/reg.html accessed April 29, 2003 for information on the Mindsong RNG, which is no longer manufactured.

14. Specifically, checks are run against the expected mean, variance, skew, and kurtosis for truly random binomial samples.

15. The RNGs were programmed to continuously generate samples, each consisting of 200 random bits. Use of summed samples, rather than individual random bits, is intentional as it diminishes potential biasing effects due to short-term autocorrelations that may arise in the bit sequences. In addition, the raw bit sequence was run through an exclusive-or (XOR) logic gate against a pattern of an equal number of 0 and 1 bits to guarantee that the mean output is unbiased. These samples were collected and stored at a rate of about one sample per second, and each sample was automatically date and time-stamped.

16. Model RM-60, made by Aware Electronics, http://www.aw-el.com/ accessed February 2005.

17. Each sample from each of the two electronic noise-based RNGs was transformed into a normalized score as $z = (x-100) / \sqrt{50}$, where x was the per-second RNG sample. Then a one-hour composite Stouffer Z score, z_H, was created by combining the per-second z scores as $z_H = \Sigma z_i / \sqrt{N}$, where i ranged from 1 to 3,600 and $N = 3,600$ for the Orion RNG and i ranged from 1 to 3,970 and $N = 3,970$ for the MindSong RNG. Data from the radiation monitor were normalized by finding the mean m and standard deviation s of all 10-second counts, forming $z' = (x-m)/s$ for each sample x, and then creating a one-hour composite Stouffer Z score, z_H, as above, except with $N = 360$ (because x was based on 10-second rather than 1-second counts).

18. Spearman $R = 0.75$, $p = 0.0009$, excluding four colony counts on which the two analysts disagreed.

19. Spearman $R = 0.47$, $p = 0.07$, excluding four colony counts.

20. The order of the observed z_H values were randomly shuffled, a new 12-hour sliding window curve was formed out of the shuffled order, and then the peak value within the 51 hours of the experimental period was determined. This procedure was then repeated 100,000 times to build up a distribution of possible peak values. This analysis

showed that among the maximum shuffled z_H values only 9 exceeded the observed $z = 4.8$, thus the probability of the observed peak deviation was 9/100,000 or $p = 0.00009$.

21. Radin, D. I., Taft, R., and Yount, G. (2004). Possible effects of healing intention on cell cultures and truly random events. *Journal of Alternative and Complementary Medicine*, **10**, 103–112.

22. $r = -0.60$, $N = 39$, $p = 2.7 \times 10^{-5}$. The results still showed a significant decline when the three RNGs in the lab were excluded, $r = -0.41$, $N = 36$, $p = 0.005$.

23. Nelson, R., Boesch, H., Boller, E., Dobyns, Y., Houtkooper, J., Lettieri, A., Radin, D., Russek, L., Schwartz, G., and Wesch, J. (1998). Global resonance of consciousness: Princess Diana and Mother Teresa. *The Electronic Journal of Parapsychology*. Abstract available on http://noosphere.princeton.edu as of January 2005.

24. The method of analysis was strictly defined before examining the data.

25. The ordinate on this graph is the normalized average absolute deviation of the variance, evaluated across all GCP RNGs per second. The GCP data were extracted with respect to the stroke of Y2K midnight in 27 time zones running according to UTC, and two time zones, India and Iran, running 30 minutes off the UTC clock. These 29 time zones include an estimated 98% of the world's population. To help visualize the results, these data are smoothed with a five-minute sliding window.

26. Evaluated using a randomized permutation analysis.

27. This analysis is based on a five-minute sliding average absolute deviation from the empirical mean in 29 time zones, in which the vast majority of the world's population lives. There are actually a total of 38 time zones.

28. There is evidence associated with other sudden events of global interest, like massive earthquakes, that similar "precursor" waves appear in the global randomness network about two to three hours in advance.

29. Radin, D. I. (2002). Exploring relationships between random physical events and mass human attention: Asking for whom the bell tolls. *Journal of Scientific Exploration*, **16** (4), 533–548; Radin, D. I. (2003). For whom the bell tolls: A question of global consciousness, *Noetic Sciences Review*, **63**, 8–13 and 44–45.

30. www.infoplease.com as of December 2004; Information Please is part of Pearson Education, owner of publishing companies including Prentice Hall, Scott Foresman, Addison Wesley Longman, the Financial Times, and Penguin Putnam.

31. $r = 0.16$, t (363 df) $= 3.08$, $p = 0.001$, one-tailed; after removing 9/11 the correlation remained significantly positive, $r = 0.15$, t (362 df) $=$ 2.88, $p = .002$, one-tailed. If all of the non-news days were removed, the correlation remained significant, $r = 0.11$, t(248 df) $= 1.76$, $p =$ .040, one-tailed. I then extended this analysis to all events from 1998 through 2002, and the correlation still remained significant ($p = 0.002$).

32. Nelson, R. D., Radin, D. I., Shoup, R., and Bancel, P. (2002). Correlation of continuous random data with major world events. *Foundations of Physics Letters*, **15** (6), 537–550.

Chapter 12: A New Reality

1. Evidence for mind-matter interaction in metal-bending is discussed in detail in Hasted, J. (1981). *The Metal-benders*. London: Routledge & Kegan Paul. I was much more skeptical about such claims until one day I personally folded the bowl of a large, heavy soup spoon in half with a gentle touch, and with half a dozen witnesses present. I later tested to see if I could do this again with a similar spoon using ordinary force. I couldn't budge the bowl without the assistance of two pairs of pliers and some serious leverage. So I have good reason to doubt the usual skeptical assertion that all cases of metal-bending are conjuring tricks or due to unconscious use of force.

2. I'm aware that there is some serious evidence for UFOs and Bigfoot and other unorthodox claims. The point is that the *type* of evidence for such claims differs considerably from the controlled laboratory studies we've been discussing.

3. This isn't the precise technical definition of locality, but it's close enough for our purposes.

4. William Thomson (Lord Kelvin), in his lecture given at the Royal Institute "Nineteenth century clouds over the dynamic theory of heat and light." Published originally in *Philosophical Magazine,* **2**, 1–40, July 1901.

5. Some today believe that positive evidence for an ether was obtained by physicist Dayton Miller in the early part of the twentieth century, and reported in a 1933 paper in *Reviews of Modern Physics*. But also see a review of tests of Einstein's theory of special relativity in *Science,* February 11, 2005.

6. Niels Bohr considered complementary states so fundamental that he also applied the concept to philosophical and psychological concepts.

This included efficient and final cause, conscious and unconscious states, thinking and feeling, goodness and justice, confirmation and novelty, mind and matter. The nonlocal entanglements inherent in such complementary states were thought by psychiatrist Carl Jung and physicist Wolfgang Pauli to give rise to meaningful synchronicities.

7. Feynman, R. (1990). *QED, The Strange Theory of Light and Matter,* Penguin Books, London, p. 9.
8. http://physicsweb.org/articles/world/15/9/1.
9. The confusion is at least partially due to our thinking that a photon is like an everyday object. It's much more mysterious.
10. Aharonov, Y. and Zubairy, M. S. (2005). Time and the quantum: Erasing the past and impacting the future. *Science,* **307**, 875–879.
11. This is accomplished by erasing "which-path" information. See Aharonov and Zubairy, ibid., for details.
12. Rosenblum, B., and Kuttner, F. (2002). The observer in the quantum experiment. Department of physics, University of California Santa Cruz. *Foundations of Physics,* **32** (8), 1273–1293.
13. Not all physicists would agree with this. See, for example, May, E. C., Spottiswoode, S. J. P., and Piantanida, T. (1988). Testing Schrodinger's paradox with a Michelson interferometer. *Physica B,* **151**, 339–348.
14. Greene, B. (2004). *The Fabric of the Cosmos,* New York: Alfred A. Knopf, p. 119.
15. Mermin, N. David (1990). *Boojums All the Way Through: Communicating Science in a Prosaic Age.* Cambridge, UK: Cambridge University Press, p. 119.
16. The concept of quantum superposition refers to the wave aspect of quantum objects. It is based on the mathematical description of an unobserved quantum object, which is said to exist in a probabilistic, wavelike mixture of possible states rather than a single actual state.
17. The key words here are "ordinary assumptions." In everyday life we never observe anything with "indefinite properties," so it's hard to imagine what that might be like.
18. Mermin, N. D. (April 1985). Is the moon there when nobody looks? Reality and the quantum theory. *Physics Today,* 38–47.
19. Bohr, Niels (1963). *Essays 1958/1962 on Atomic Physics and Human Knowledge.* New York: Wiley, p. 15.
20. Goswami, A., Reed, R. E., and Goswami, M. (1995). *The Self-Aware Universe: How Consciousness Creates the Material World,* New York: Jeremy Tarcher/Putnam.

21. Greene (2004), p. 212, op. cit., italics in the original.
22. Bub, J. (1997). *Interpreting the Quantum World*. Cambridge, UK: Cambridge University Press.
23. Clarke, C. (Winter 2004). *Network*, no. 86, Scientific and Medical Network, 13–16.
24. Einstein, A., Podolsky, B., and Rosen, N. (1935). Can a quantum mechanical description of physical reality be considered complete? *Physical Review,* **47**, 777–780.
25. Born, M., and Einstein, A. (2005). *The Born-Einstein Letters: Friendship, Politics and Physics in Uncertain Times*. New York: Macmillan. "In an age of mediocrity and moral pygmies, their lives shine with an intense beauty. Something of this is reflected in their correspondence and the world is richer for its publication."–Bertrand Russell, from original Foreword.
26. Aczel, 2001, op. cit., p. 235.
27. Ibid., p. 70.
28. Cited on PhysicsWeb, online version of *Physics World* magazine, December 1998; http://physicsweb.org/articles/world/11/12/8 original from Henry Stapp (1977). *Nuovo Cimento,* **40B**, 191.
29. Physicist John F. Clauser, while at Columbia University in 1969, thought of a way to test Bell's theorem using polarizers. In 1972, at the University of California at Berkeley, Stuart Freedman and Clauser published their experimental results, which was in accordance with quantum theory; Freedman, S. and Clauser, J. (1972). Experimental test of local hidden variable theories. *Physical Review Letters,* **28**, 934–941. In 1973, Holt and Pipkin replicated the effect with mercury atoms. This was followed in 1976 by Clauser, and by Edward Fry and Randall Thompson at Texas A&M; in 1974 by Faraci, Gutkowski, Notamigo, and Pennisi at the University of Catania, Italy; and in 1976 by Lamehi-Rachti and Mittig at the Saclay Nuclear Research Center near Paris.
30. Marcikic, I., de Riedmatten, H., Tittel, W., Zbinden, H., Legré, M., and Gisin, N. (2004). Distribution of time-bin entangled qubits over 50 km of optical fiber. *Physical Review Letters,* 93; Tittel, W., Brendel, J., Gisin, B., Herzog, T., Zbinden, H., and Gisin, N. (1998). Experimental demonstration of quantum correlations over more than 10 km. *Physical Review A,* **57**.
31. Aczel, 2001, op. cit., p. 203.
32. Aczel, 2001, op. cit., p. 249.
33. Mermin. N. D. (1985). Is the moon there when nobody looks? Reality and the quantum theory. *Physics Today,* April 1985, 38–47.

34. Marcikic et al., 2004, op. cit.
35. Ibid.
36. Greene (2004), p. 84, op. cit.
37. Greene (2004), p. 113, op. cit.
38. James, W. (October 1909). The final impressions of a psychical researcher. *The American Magazine,* Reprinted in Gardner Murphy and Robert O. Ballou (eds.), *William James on Psychical Research.* London: Chatto and Windus, 1961.
39. Mauskopf and McVaugh, 1980, op. cit, p. 61.
40. Jahn, R. G., and Dunne, B. J. (2001). A modular model of mind/matter manifestations (M5). *Journal of Scientific Exploration,* vol. 15, no. 3, pp. 299–329.
41. James, W. (1897). *The Will to Believe.* New York: Longmans, Green & Co.
42. Note that he did not say "supernatural," which implies beyond nature, or divine.
43. James, W. (October 1909), op. cit.
44. A similar concept is discussed by general systems theorist Ervin Laszlo in his 1995 book, *The Interconnected Universe: Conceptual Foundations of Transdisciplinary Unified Theory,* New Jersey: World Scientific; and in his 2003 book, *The Connectivity Hypothesis: Foundations of an Integral Science of Quantum, Cosmos, Life and Consciousness.* Albany, NY: State University of New York Press.
45. Personal communication, December 25, 2004.
46. By convention the names used in providing human-scale descriptions of quantum experiments are Alice and Bob. Here I retain Jack and Jill to underscore the similarity with the results of psi experiments.
47. The ideal theoretical outcome when the photos were the same would be 100%. I chose 77% to provide a more realistic outcome, and also so the overall hit rate would be at 59% and match the result of the dream psi studies.
48. http://inexistant.net/Gilles/en/ as of December 29, 2004.
49. Vandegrift, G. (1995). Bell's theorem and psychic phenomena. *The Philosophical Quarterly,* 45 (181), 471–476.
50. Ibid., p. 476.
51. A third reaction comes from physicists who are so used to the ideas of quantum weirdness that they no longer understand what all the fuss is about. The commotion revolves around the ontological and epistemological challenges of quantum theory more than any pragmatic conse-

quences. At least for now. When basic worldviews shift, all sorts of changes are catalyzed.

52. See Endnote 48.

Chapter 13: Theories of Psi

1. Schilpp, P. A. (1949). *Albert Einstein: Philosopher-Scientist*. La Salle, IL: Open Court, p. 683.

2. Stokes, D. M. (1987). Theoretical parapsychology. In Krippner, S. (ed.), *Advances in Parapsychological Research 5* (pp. 77–189). Jefferson, NC: McFarland.

3. Rauscher, E. A., and Targ, R. (2001). The speed of thought: Investigation of a complex space-time metric to describe psychic phenomena. *Journal of Scientific Exploration,* **15** (3), 331–354.

4. Becker, R. O. (1992). Electromagnetism and psi phenomena. *Journal of the American Society for Psychical Research,* **86** (1): 1–17; Persinger, M. A. (1989). Psi phenomena and temporal lobe activity: The geomagnetic factor. *Research in Parapsychology,* Metuchen, NJ: Scarecrow Press: 121–156; Persinger, M. A. and Krippner, S. (1989). Dream ESP experiences and geomagnetic activity. *Journal of the American Society for Psychical Research,* **83**, 101–116.

5. There is some evidence that complexity might matter in precognition. E.g., Vassy, Z. (1986). Experimental study of complexity dependence in precognition. *Journal of Parapsychology,* **50**, 235–270.

6. Schmidt, H. (1975). Towards a mathematical theory of psi. *Journal of the American Society for Psychical Research,* **69** (4), 301–320.

7. Stanford, R. G. (1990). An experimentally testable model for spontaneous psi events. In Krippner, S. (ed.), *Advances in Parapsychological Research 6,* McFarland and Co., pp. 54–167; Stanford, R. G., Zenhausern, Z., Taylor, A., and Dwyer, M. (1975). Psychokinesis as psi mediated instrumental response. *Journal of the American Society for Psychical Research,* **69** (2), 127–134 .

8. Thalbourne, M. A. (2005). The theory of psychopraxia: A paradigm for the future? In M. Thalbourne and L. Storm (eds.), *Parapsychology in the Twenty-first Century.* Jefferson, NC: McFarland & Company, pp. 189–204.

9. May, E. C., Utts, J. M. and Spottiswoode, S. J. P. (1995). Decision Augmentation Theory: Towards a model of anomalous phenomena. *Journal of Parapsychology,* **59** (3), 195–220; May, E. C., Spottiswoode, S. J. P., Utts, J. M., and James, C. L. (1995). Applications of Decision Augmentation Theory. *Journal of Parapsychology,* **59** (3), 221–250; Dobyns,

Y. H. (1993). Selection versus influence in remote REG anomalies. *Journal of Scientific Exploration,* **7** (3), 259–269; Dobyns, Y. H. (1996). Selection versus influence revisited: New methods and conclusions. *Journal of Scientific Exploration,* **10** (2), 253–268.

10. Persinger, M. A., Roll, W. G., Tiller, S. G., Koren, S. A., and Cook, C. M. (2002). Remote viewing with the artist Ingo Swann: Neuropsychological profile, electroencephalographic correlates, magnetic resonance imaging (MRI), and possible mechanisms. *Perceptual and Motor Skills,* **94**, 927–949.

11. Sheldrake, R. (1992). An experimental test of the hypothesis of formative causation. *Rivista di Biologia–Biology Forum,* **86** (3/4), 431–444.

12. Hardy, C. (2000). Psi as a multilevel process: Semantic fields theory. *Journal of Parapsychology,* **64**, 73–94.

13. Stokes, 1987, op. cit., p. 156–163.

14. Smith, W. W. (1920). *A Theory of the Mechanism of Survival: The Fourth Dimension and Its Applications.* New York: E. P. Dutton and Co.

15. Stapp, H. (in press). Quantum Approaches to Consciousness. In Moscovitch, M., and Zelazo, P. (eds.). *Cambridge Handbook for Consciousness.*

16. Stokes, 1987, op. cit., p. 172; Penfield, W. (1975). *The Mystery of the Mind: A Critical Study of Consciousness and the Human Brain.* Princeton, NJ: Princeton University Press.

17. Wigner, E. P. (1967). *Symmetries and Reflections.* Cambridge, Mass: MIT Press, p.171–184.

18. A phrase attributed to Nobel Laureate Murray Gell-Mann.

19. d'Espagnat, B. (November 1979). The quantum theory and reality. *Scientific American,* 158–181.

20. Walker, E. H. (2000). *The Physics of Consciousness.* Cambridge, MA: Perseus Books.

21. Walker, E. H. (1975). Foundations of paraphysical and parapsychological phenomena. In Oteri, L. (ed.) *Quantum Physics and Parapsychology,* Parapsychology Foundation; Walker, E. H. (1984). A review of criticisms of the quantum mechanical theory of psi phenomena, *Journal of Parapsychology,* **48**, 277–332; Schmidt, H. (1984). Comparison of a teleological model with a quantum collapse model of psi, *Journal of Parapsychology,* **48** (4), 261–276; Schmidt, H. (1975). Toward a mathematical theory of psi. *Journal of the American Society for Psychical Research,* **69**, 301–319.; Walker, E. H. (1973). Application of the quantum theory of consciousness to the problem of psi phenomena. In Roll, W. G., Morris, R. L., and Morris, J. D. (eds.), *Research in Parapsychology* 1972 (pp.

51–53). Metuchen, NJ: Scarecrow Press; Walker, E. H. (1975). Foundations of paraphysical and parapsychological phenomena. In Oteri, L. (ed.), *Quantum Physics and Parapsychology* (pp. 1–44). New York: Parapsychology Foundation.

22. Houtkooper, J. M. (2002). Arguing for an observational theory of paranormal phenomena. *Journal of Scientific Exploration,* **16** (2), 171–185.

23. Bierman, D. J. (1998). Do psi phenomena suggest radical dualism? In: Stuart R. Hameroff, Alref W. Kaszniak, and Alwyn C. Scott (eds.), *Toward a Science of Consciousness II,* Cambridge MA, MIT Press: pp. 709–714; There have been related tests examining whether the person who analyzes experimental data influences the outcome, the so-called checker or analyst effect. These studies have had positive outcomes, supporting the retro-PK prediction.

24. Von Lucadou, W. (2001). Hans in luck: The currency of evidence in parapsychology. *Journal of Parapsychology,* **65,** 3–16; von Lucadou, W. (1991). The model of pragmatic information (MPI), *European Journal of Parapsychology,* **11,** 58–75.

25. Walach, H. (2003). Generalized entanglement: Possible examples, empirical evidence, experimental tests. In Proceedings: *Generalized Entanglement from a Multidisciplinary Perspective,* eds. Rainer Schneider, Ronald Chez, Freiburg, Germany, October 2003, p. 66–95.

26. Jahn, R. (1991). The complementarity of consciousness. PEAR technical report, PEAR 91006, December 1991. Available from http://www.princeton.edu/~pear/publist.html as well as Jahn, R., and Dunne, B. (1987). *Margins of Reality: The Role of Consciousness in the Physical World.* New York: Harcourt Brace Jovanovich.

27. Atmanspacher, H., Romer, H., and Walach. H. (2002). Weak quantum theory: Complementarity and entanglement in physics and beyond. *Foundations of Physics,* **32** (3), 379–406.

28. Bohm, D. (1980). *Wholeness and the Implicate Order.* London: Routledge Classics.

29. Ibid.

30. Transcript of an interview with Jeffrey Mishlove from the television show, *Thinking Allowed.*

31. Wilber, K. (1982). *The Holographic Paradigm and Other Paradoxes.* Boulder, CO: Shambhala; Wilber, K. (1984). *Quantum Questions.* Boulder, CO: Shambhala. Also see Roney-Dougal, S. (1993). *Where Science and Magic Meet.* Rockport, MA: Element Books.

32. Bekenstein, J. D. (February 2004). Information in the Holographic Universe. *Scientific American* (Online), 29–36.

33. Marcer, P. J., and Schempp, W. (1997). Model of the neuron working by quantum holography, *Informatica,* **21**, 1997, 519–534; Marcer, P. J., and Schempp, W. (1998). The brain as a conscious system. *International Journal of General Systems*; Marcer, P., Mitchell, E., and Schempp, W. (2002). Self-reference, the dimensionality and scale of quantum mechanical effects, critical phenomena, and qualia. *International Journal of Computing Anticipatory Systems,* **13**, 340–359.

34. Schewe, P., Riordon, J., and Stein, B. (2001). American Institutes of Physics, Number 566 #1, November 21, 2001, from http://www .aip.org/pnu/2001/split/566-1.html accessed as of January 2005.

35. Stapp, H. (2004). Physics in neuroscience. In M. Beauregard (ed.). *Consciousness, Emotional Self-regulation, and the Brain: Advances in Consciousness Research Series #54,* Amsterdam/Philadelphia: John Benjamin Books.

36. Von Neumann, John (1955). *Mathematical Foundations of Quantum Theory.* Princeton: Princeton University Press.

37. Stapp, 2004, op. cit.

38. Hagan, S., Hameroff, S. R., and Tuszynski, J. A. (2002). Quantum computation in brain microtubules: Decoherence and biological feasibility. *Physical Review E,* **65**.

39. Zeno refers to the Greek philosopher, Zeno of Elea, who proposed the famous Zeno Paradox: Imagine an arrow flying through the air. If the motion of this arrow is subdivided into an infinite number of infinitely small points, then at each of those points the arrow will have zero velocity. Since the sum of an infinite number of zeros is still zero, we must conclude that the arrow has no velocity, and is not actually moving. The paradox, of course, is that in spite of our impeccable logic the arrow will still hit the target.

40. A similar proposal is made by biophysicist Johnjoe McFadden in his 2002 book, *Quantum Evolution: How Physics' Weirdest Theory Explains Life's Biggest Mystery,* New York: W. W. Norton & Company.

41. Nadeau, R., and Kafatos, M. (2001). *The Non-local Universe: The New Physics and Matters of the Mind.* Oxford University Press, p. 216.

42. Nadeau and Kafatos, 2001, op. cit., p. 81, italics added for emphasis.

43. Ibid., p. 100.

44. Bergson, H. (1914). Presidential address. *Proceedings of the Society for Psychical Research,* **27**, 157–175; Huxley, A. (1954) *The Doors of Perception.* London: Chatto & Windus.

45. Bergson (1914), op. cit.

46. Greene (2004), op. cit., p. 122.

47. Summhammer (2005), op. cit.

48. Perhaps some people who complain about unwanted thoughts appearing in their heads aren't hallucinating. The problem is that being supersensitive to genuine psi might well drive one crazy, and thus clearly distinguishing between psi and psychopathology becomes difficult.

Chapter 14: Next

1. This overall figure is based on the probability associated with an unweighted Stouffer Z across the seven classes of experiments. This figure is undoubtedly conservative because it excludes other classes of psi experiments that weren't discussed in this book.
2. Rinpoche, S. (1992). *The Tibetan Book of Living and Dying*. San Francisco: HarperCollins; Wangval, T. (1998). *Tibetan Yogas of Dream and Sleep*. Ithaca, NY: Snow Lion Publications.
3. This assumes $50 million divided by $100 billion = 0.05%, and thus 0.05% of 24 hours = 43 seconds. A hundred billion spent on cancer research is probably a conservative estimate.
4. Kaiser, J. (2005). NIH chief clamps down on consulting and stock ownership. *Science,* **307**, 824.
5. Sidgwick, H. (1882). Presidential address. *Journal of the Society for Psychical Research,* **1**, p. 8, 12
6. Stenhoff, M. (2002), *Ball Lightning—An Unsolved Problem in Atmospheric Physics.* New York: Kluwer Academic.
7. Turner, D. J. (2002). The fragmented science of ball lightning (with comment); *Philosophical Transactions of the Royal Society,* London A, **360**, p. 108.
8. Abrahamson, J., and Dinniss, J. (2000). Ball lightning caused by oxidation of nanoparticle networks from normal lightning strikes on soil. *Nature,* **403**, 519–521.
9. Turner, D. J., (1994). The structure and stability of ball lightning. *Philosophical Transactions: Physical Sciences and Engineering,* **347** (1682), 83–111.
10. Alcock, J. E. (2003). Give the null hypothesis a chance: Reasons to remain doubtful about the existence of psi. In Alcock, J. E., Burns, J. E., and Freeman, A. (eds.), *Psi Wars: Getting to Grips with the Paranormal.* Charlottesville, VA: Imprint Academic, p. 29–50.
11. Jeffers, S. and Sloan, J. (1992). A low light level diffraction experiments for anomalies research. *Journal of Scientific Exploration,* **6**, 333.
12. Jeffers, S. (2003). Physics and claims for anomalous effects related to consciousness, 135–152. In Alcock, J. E., Burns, J. E., and Freeman,

A. (eds.), *Psi Wars: Getting to Grips with the Paranormal*. Charlottesville, VA: Imprint Academic.

13. But when the PEAR Lab tried to replicate Jeffers's experiment using his equipment, they did report a significant outcome. Ibison, M. and Jeffers, S. (1998). A double-slit diffraction experiment to investigate claims of consciousness-related anomalies. *Journal of Scientific Exploration*, **12**, 543–550.

14. Freedman, M., Jeffers, S., Saeger, K., Binns, M., and Black, S. (2003). Effects of frontal lobe lesions on intentionality and random physical phenomena, *Journal of Scientific Exploration*, **17**, 651–668.

15. Website of Psi Chi, the national honor society in psychology: http://www.psichi.org/pubs/articles/article_121.asp as of January 2005.

16. Haw, R. M. and Fisher, R. P. (2004). Effects of administrator-witness contact on eyewitness identification accuracy. *Journal of Applied Psychology*, **89** (6), 1106–1112; Good, T. L. and Nichols, S. L. (2001). Expectancy effects in the classroom: A Special focus on improving the reading performance of minority students in first grade classrooms. *Educational Psychologist*, **36** (2), 113–126; Rosenthal, R. (2002). Covert communication in classrooms, clinics, courtrooms, and cubicles. *American Psychologist*, **57** (11), 839–849.

17. Website of Psi Chi, the national honor society in psychology, http://www.psichi.org/pubs/articles/article_121.asp as of January 2005.

18. Nicole, M. K., and Kierein, N. M. (2000). Pygmalion in work organizations: a meta-analysis. *Journal of Organizational Behavior*, **21**, 913–928; Rosenthal, R. (2003). Covert communication in laboratories, classrooms, and the truly real world. *Current Directions in Psychological Science*, **12** (5), 151–154.

19. Gandar, J. M., Zuber, R. A., and Lamb, R. P. (2001). The home field advantage revisited: a search for the bias in other sports betting markets, *Journal of Economics and Business*, **53**, 439–453.

20. Vergin, R. C., and Sosika, J. J. (1999). No place like home: an examination of the home field advantage in gambling strategies in NFL football. *Journal of Economics and Business*, **51**, 21–31.

21. Camp, B. H. (1937). Statement in Notes Section. *Journal of Parapsychology*, **1**, 305

22. McClenon, J., Roig, M., Smith, M. D., and Ferrier, G. (2003). The coverage of parapsychology in introductory psychology textbooks: 1990–2002. *Journal of Parapsychology*, **67**, 167–179.

23. These are headlines from the *Weekly World News*.

24. Ted Dace, http://www.skepticalinvestigations.org/exam/Dace_amazing3
.htm as of January 2005.

25. The high jump world record has been held by Javier Sotomayer from
Cuba since 1989, as of January 2005.

26. McMoneagle, J. (2002). *The Stargate Chronicles: Memoirs of a Psychic Spy.*
Charlottesville, VA: Hampton Roads.

27. This is reproduced from one report out of tens of thousands of documents produced by the U.S. government's 20-year program on psi research and applications. A portion of these documents are now
available from the CIA through the Freedom of Information Act.

28. There are important meta-questions associated with psi that I've
avoided discussing here. Genuine psi presents serious epistemological
and ontological challenges that will need to be clarified before ro-
bustly reliable psi-based technologies can be devised. Some elementary psi technologies can be produced with today's knowledge, but
they won't be very reliable.

29. Morgan, K., and Morris, R. (1991). A review of apparently successful
methods for the enhancement of anomalous phenomena. *Journal of the
Society for Psychical Research,* **58,** 1–9; Radin, D. I. (1990). Testing the
plausibility of psi-mediated computer system failures. *Journal of Parapsychology,* **54,** 1–19; Radin, D. I. (1990–1991). Statistically enhancing
psi effects with sequential analysis: A replication and extension. *European Journal of Parapsychology,* **8,** 98–111; Radin, D. I. (1993). Neural
network analyses of consciousness-related patterns in random sequences. *Journal of Scientific Exploration,* **7** (4), 355–374.; Radin, D. I.
(1996). Towards a complex systems model of psi performance. *Subtle
Energies and Energy Medicine,* **7,** 35–70.

30. Surowiecki, J. (2004). *The Wisdom of Crowds.* New York: Doubleday.

INDEX

absorption, 41
"Age of Industry," 241–42
"Age of Information," 242–43
"Age of Magic," 241
alchemy, 243
Alcock, James, 283
Allen, Paul, 169
altered states, 269–70
Alternative Therapies in Health and Medicine, 137
American Psychologist, 77
animal magnetism, 60–61
anima mundi, 241
antimatter, 247
apparitions, 208
apples-and-oranges problem, 103
Aristotle, 57, 58, 247
Armageddon, prophecies of, 199
Aspect, Alain, 76, 226–27
Aston, Francis, 67–68
astral bodies, 243
astrocytes, 186
astrology, 243
Atmanspacher, Harald, 253
atomic bomb, 211, 268
atomic structure, 68
attention, 260, 264
autocorrelation, 206
autonomic nervous system, 131–33

Bacon, Sir Francis, 57–59
ball lightning, 280–82

Bancel, Peter, 206
Bardeen, John, 11
bardos, 277
Barrett, Sir William, 63
Beautiful Mind, A (film), 50
Behavioral and Brain Science, 77
Behavioral Neuroscience, 137
behaviorism, 242
belief, 35–51
    and blindness to the obvious, 43–44
    characteristics of believers, 44–46
    and education level, 39–40
    and expectation, 43
Bell, John, 75, 226
bell curve, 202
Bell Laboratories, 72
Bell's inequality, 230–31
Bell's theorem, 75, 76, 226, 227–31, 238
Bem, Daryl, 117–18, 119
Bender, Hans, 72
Berger, Hans, 21–24
Berger rhythms, 23
Bergson, Henri, 267
Besant, Annie, 68
Bial Foundation, 78
Bierman, Dick, 78, 176–79
bioentanglement, 16
Blackmore, Susan, 48
Blake, William, 263
blindness, inattentional, 43–44

bobbing brain idea, 266
Bohm, David, 254–55
Bohm's implicate/explicate order,
 254–56
Bohr, Niels, 213, 222, 253, 262
BOLD (Blood Oxygenation Level
 Dependent), 177
Born, Max, 68
Bousso, Raphael, 13
Boyden, Jim, 169
brain:
 astrocytes in, 186
 bobbing, 266
 clockworks of, 241–42, 257
 communication infrastructure
  of, 258
 EM fields generated by, 246
 as filter, 267
 and holograms, 255
 and the mind, 73, 243, 260,
  264, 267
 misfiring, 47–49
 neurons in, 13
 and religion, 47–48
 responding to future events,
  179
 role of quantum theory in, 243
 self-observed, 260
Brassard, Gilles, 237, 238
Braud, William, 117
British Journal of Psychology, 40, 134
Broglie, Louis de, 213
Brooks, Michael, 14
Broughton, Richard, 173–74
Brugmans, H. I. F. W., 82
Bub, Jeffrey, 225

Caesar, Julius, 147
Camp, Burton, 288
Carington, Whately, 72, 93–95,
 250
Carlson, Chester, 69
Carson, Shelley, 49, 50
Caryl, Peter, 138–39

causality:
 in classical physics, 210
 in new reality, 221
 and uncertainty principle,
  214–15
causes, efficient and final, 247–48
Central Intelligence Agency (CIA),
 76–77, 78, 99
central nervous system, 136–41
certainty, 221
chaos theory, 11
Child, Irvin, 77
Cicero, 57
clairvoyance, 6, 68, 81, 125
 experiment in, 70–72
 Pearce-Pratt test, 86–89
 picture-drawing test, 89–95
 and precognition, 162
 traveling, 48
Clarke, Chris, 225
classical physics, 209–10, 257
 basic assumptions about reality
  in, 210, 221
 changing assumptions of,
  220–21
 field theories in, 243–44
 objective reality in, 218
classical science era, 241–42
Clauser, John, 227
clockworks, 2, 241–42, 257
coherence, 2, 16
coherence effects, 185
coincidence, frequency of, 245
coin flipper, electronic, 154–56
cold fusion, 13
collective unconscious, 248
common sense:
 and belief, 42–43
 either-or logic of, 223–24
 in new reality, 220
 presentiment vs., 168
complementarity, 214, 220,
 253–54
conditioned-space hypothesis, 189

confirmation bias, 102
conformance theory, 248
conscious awareness, 2, 235–36, 257, 265
consciousness:
    altered states of, 269–70
    collective forms of, 270
    cosmic, 249
    in fourth dimension, 250
    nature of, 277
    normal, 234
    quantum approaches to, 250–51
    and reality, 224
    sustaining, 243
    understanding, 236
contingent negative variation (CNV), 163
continuity, 210, 221
Coover, John Edgar, 67
Copenhagen interpretation, 222–23, 224, 256
correlation, 185, 264
cosmology, 12
creativity, 50–51
critical thinking ability, 40
Crookes, Sir William, 62
Curie, Marie and Pierre, 66
Custer, George, 63
cybernetics, 11, 243

Dace, Ted, 290
Dag, I., 41
Dalton, Kathy, 119
dark energy and matter, 12
Dawkins, Richard, 290
Dean, Douglas, 106
death, transition to, 277
decision augmentation theory, 248
decline effects, 88, 121–25, 163, 268–69
decoherence, 16, 224–25, 258–59
Delphic oracle, 55–57
delusion, 246

Democritus, 57
Descartes, René, 24, 209
d'Espagnat, Bernard, 251
determinism, 210, 221
Diana, Princess, 193–94
dice-tossing experiments, 148–53
Dick, Philip K., 24–25
displacement, 268, 269
distance-dependence properties, 191–93
distant information, capacity to receive, 180
distant staring, 133–34
DMILS (direct mental interactions with living systems), 131–33
    distant staring in, 133–34
    meta-analysis, 134–36
DNA, junk, 12–13
Donne, John, 207
double-slit experiment, 215–20
dream incubation, 55
dreams, psi in, 104–9, 111, 115, 269–70
dream telepathy, 75
Drühl, Kai, 217
Dunne, Brenda, 95, 96, 232–33, 253

earthworm experiments, 170–71
Eccles, Sir John, 73–74, 251
Edgeworth, F. Y., 64
efficient and final cause, 247–48
Einstein, Albert, 254
    and classical physics, 210
    on entanglement, 14, 72
    and EPR paper, 214, 220, 225–26
    and neorealism, 225
    on photoelectric effect, 67
    and quantum theory, 14, 72, 213, 214, 220–21, 225, 227, 231
    Special Theory of Relativity, 268

Einstein, Albert *(cont.)*
  on spooky action at a distance,
    1, 14, 221, 226
  on telepathy, 67, 91, 240
electroencephalogram (EEG), 23,
  136–37, 138–39
electrogastrogram (EGG), 143
ELF waves, 247
Elworthy, Frederick Thomas, 126
embellishment, 245
emotions-at-a-distance, 143
empirical reasoning, 58
energy, relative, 242
entangled brains experiments, 18
entangled minds, 260–62
  assumptions of, 269
  and field-consciousness effects,
    270
  inside, 263–66
  questions about, 266–67
entanglement:
  and coherence, 2, 16
  coining of term, 226
  confirmation of, 227
  correlation in, 185, 264
  demonstrations of, 15–16
  in disciplines beyond physics,
    238–39
  in the future, 17–19
  generalized, 253
  photons entangled with
    themselves, 217
  in quantum theory, 1–2, 14, 72,
    76, 220, 221, 226, 235,
    237–38
  theories of, 244
EPR paper, 214, 220, 225–26
error hypothesis, 241
Ertel, Suitbert, 118
ESP: A Scientific Evaluation
  (Hansel), 88–89
ESP cards, 72, 83–89, 162, 193,
  288, 289
ESPER lab, 78

Everett, Hugh, 223
evil eye, 125–26
expectations, 43, 186, 285
experimenter expectancy effect,
  285–87
Experiments in Mental Suggestion
  (Vasiliev), 74
explicate order, 254–55
extrasensory perception (ESP):
  automated testing machines for,
    68, 70
  ESP card tests, 83–89, 162,
    193, 288, 289
  research in, 74, 76, 82
Extra-Sensory Perception (Rhine), 70
Extrasensory Perception After Sixty Years
  (Rhine), 72, 84–85
eyewitness testimony, 246

fascination, 125
Fayed, Dodi al-, 193–94
Fechner, Gustav Theodor, 62–63
Ferrari, Diane, 148, 161–62
Feynman, Richard, 215, 217
field-consciousness experiments,
  182–85, 191, 194, 205, 207,
  270
field theories, 243–44, 248–49
Figar, Stepán, 74
file-drawer problem, 103, 104, 112
final cause, 247–48
Fisher, Sir R. A., 68
focused group energy, 183–84
Foerster, Heinz von, 11
forced-choice tests, 161–62
Foundation of Physics Letters, 206
Foundations of Physics, 253
fourth dimension, 249–50
Fox, Margaretta and Catherine, 61
Franklin, Benjamin, 60
fraud, 246
  participant, 115
"Freeze Frame" (self-regulation
  training program), 172

French Academy of Sciences, 60
French Royal Society of Medicine, 60–61
Freud, Sigmund, 65, 69
functional magnetic resonance imaging (fMRI), 23, 137, 176–77
funnel plot, 112–14

Galen, Claudius, 42
Galileo Galilei, 209
Galton, Sir Francis, 62
Ganzfeld studies, 115–25
   decline effects, 121–25
   meta-analysis, 120–21
   textbook discussions of, 289
Garrett, Eileen, 73, 106–7
gaze, power of, 125–30
Geller, Uri, 54
genius, and madness, 51
geomagnetic field theory, 248
Geronimo, Chief, 65, 66
Gisin, Nicholas, 227, 231
Global Consciousness Project (GCP), 195–201
   and applicability of psi, 295
   and field-consciousness effects, 205
   as human-centric, 201
   and newsworthy events, 205–7
   and pope's funeral, 197
   and RNG network, 191, 195–97
   and September 11 attacks, 202–7
   and Y2K, 199–201
goal-oriented theories of psi, 247–48, 268
Good, A. J., 163
Goswami, Amit, 224
Goulding, Anneli, 41
Granqvist, Pehr, 48
gravitons, 247
Greenberger, Daniel, 227

Greene, Brian, 219, 224, 231, 271–72
Grinberg-Zylberbaum, Jacobo, 137
gut feelings, 6, 142–45

Hameroff, Stuart, 170–71
Hansel, Mark, 88–89
Hardy, Christine, 249
Harman, Willis, 9
Hartwell, John, 164
Harvey, William, 42–43
haunted sites, 186
Hawthorne, Nathaniel, 62
healing, intentional, 185–91
heart-rate experiments, 172–73
Heisenberg, Werner, 68, 214
Herbert, Frank, 263
Herbert, Nick, 235
Herodotus, 56
Hertz, Heinrich, 65, 67
high aim condition, 154
history, importance of, 52–54
holistic concepts, 243, 244, 266
holograms:
   entire universe as, 13
   as metaphor, 254–55
holographic paradigm, 255–56
Home, Daniel Dunglas, 62
home team advantage, 287
Honorton, Charles, 76, 117–18, 161–62
Houdin, Jean Eugene Robert-, 61
hunches, 6, 142
hypnotism, 23, 60–61, 74

Ibuka, Masaru, 78
ignorance hypothesis, 36, 246
illusions, 257
implicate order, 254–55
implicit learning, 245
Innocent VIII, Pope, 57
insects, quantum cooperation of, 273
Institut d'Optique, 76, 226–27

Institute of Noetic Sciences, 76, 139–41
Institut Metapsychique International, 68
"Integral Age," 243
intention, 185, 260, 264
intentional healing, 185–91
intercorrelation value, 204, 205
Interval Research Corporation, 78, 169
intuition, 142–45
ion channels, 258
isotope, 68

Jahn, Robert:
    on generalized entanglement, 232–33, 253
    and PEAR laboratory, 76, 95, 154, 232
    and remote perception, 95, 96
    and RNG research, 154
James, William, 64, 232, 233–34, 249
James I, king of England, 57
Jeffers, Stanley, 284–85
John Paul II, Pope, 197
Johrei healing treatments, 186–91
Jordan, Pascual, 68–69, 219, 220
Josephson, Brian, 16, 224
Journal of Alternative and Complementary Medicine, 138, 172
Journal of Parapsychology, 69, 70
Journal of Personality and Social Psychology, 49
Jung, Carl, 11, 66, 131, 248
junk DNA, 12–13

Kafatos, Menas, 261–62
Kelvin, William Thomson, Lord, 211
Kennedy, James, 163
Kennedy, John F., 74
Kepler, Johannes, 209
King, Martin Luther Jr., 74

Kittenis, Marios, 138–39
Koestler, Arthur, 77
Kogan, I. M., 247
Krippner, Stanley, 107
Kroc, Ray, 74
Kuttner, Fred, 218, 219

Langley, Samuel P., 64
latent inhibition, 49–51
Leadbeater, Charles, 68
learning, implicit, 245
Levin, Jerry, 163
levitation, 208, 270
Liddle, David, 169
light:
    frequencies of, 212
    particles vs. waves of, 211–12, 213
    photoelectric effect of, 212–13
lightning, ball, 280–82
locality, 210, 221
Lodge, Sir Oliver, 64
logic, quantum, 223–24
Lourdes, spontaneous healing in, 186
low aim condition, 154
Lucadou, Walter von, 252–53
luminiferous ether, 212

MacGregor, Betsy, 25–27
madness, and genius, 51
magic, 44, 54–55, 241, 243
magnetic field effect, 48
many worlds interpretation, 223
materialism, 63
mathematics, noncommutative, 214
Mathews, Max, 169
matter, relative, 242
Maxwell, James Clerk, 210
May, Edwin, 76, 95, 101, 174–76, 248
McDonnell, James, 76
McKenna, Dennis and Terrance, 296

McMoneagle, Joseph, 292–93, 294
McRaty, Rollin, 172
measurement problem, 220
mechanical universe, 241–42
mechanism of action, 221–22
meditation, 269–70, 277
memory, tricks of, 245
mental deficiency hypothesis, 36, 41
mental intention, 58
Mental Radio (Sinclair), 70, 90–92, 246
mental telepathy, *see* telepathy
Mermin, N. David, 221, 227
Mesmer, Franz Anton, 60–61
mesmerism, 60–61
meta-analysis, 102–4, 284–85
    decline effects in, 122–23
    of DMILS, 134–36
    of dream psi studies, 109–11
    of Ganzfeld studies, 120–21
    of RNG studies, 156–58
metal bending, 208
metetherial world, 249
Michelson, Albert, 211, 212
milk production, studies of, 123
Milton, Julie, 118, 162
mind:
    and clockwork of the brain, 241–42
    cybernetic interplay of, 243
    directing the show, 260
    entangled, 260–62, 263–67, 269, 270
    functions of, 259
    influence of, 158–60, 185, 219
    in quantum-measurement process, 257
    reality affected by, 219, 251
    and soul, 241
mind-brain interaction, 73, 243, 260, 264, 267
mind-matter interaction, 19, 82, 146–60, 235
    causation in, 185

in dice experiments, 147–51
global, 197, 207
psi influence of, 146, 151–53, 158–60
and psychokinetic effects, 270–71
in RNG experiments, 154–58
spiritual source of, 63
Mitchell, Edgar, 76
model of pragmatic information, 252–53
Morley, Edward, 212
morphogenetic fields, 248
Morris, Robert, 77, 162
Mountford, Joy, 169
Mullis, Kary, 169–70
multidimensional theories of psi, 249–50
Murphy, Gardner, 232
Myers, Frederic, 66, 249
Myers-Briggs Type Indicator (MBTI), 173–74
mystics, 233

Nadeau, Robert, 261–62
Nash, John, 50
National Opinion Research Center, 38–39
National Science Foundation (NSF), 38
Nature, 48, 95, 137, 281
Nature:
    complementarity in, 253
    and conscious awareness, 235–36
    experiments about, 209–10
Nelson, Greg, 195
Nelson, Roger:
    and field-consciousness studies, 182, 194
    and Global Consciousness Project, 195, 197, 206
    and RNG studies, 156, 182, 184

NEO-Five Factor Inventory
  (NEO-FFI), 173–74
neorealism, 225–27
Neumann, John von, 224, 251,
  256–60
neurons, 13
neuroscience problem, 240
neurotheology, 57–58
neurotransmitters, 258
neutrinos, 247
Newcomb, Simon, 64
new reality, 5–6, 220–22
Newton, Sir Isaac, 209, 211
Newtonian-Cartesian worldview,
  210
Newtonian physics, 210
Niels Bohr Institute, 222
noncommutative mathematics, 214
nonlocality, 221, 261–62, 264
noosphere, 181, 243
numbers, random, 154–58

observation, reality affected by, 20,
  224, 256
Observational Theory, 251–52
occult, 210, 243
Occult Chemistry (Besant and
  Leadbeater), 68
Okada, Mokichi, 186
old wives' tales, 55
Oracle Bones, 55
oracles, 55–57
orienting reflex, 166
Ortony, Andrew, 11
Osis, Karlis, 106
out-of-body experiences, 208

Palladino, Eusapia, 66
Palmer, John, 77
paranormal, 3
  belief in, 40, 41
  and psychopathology, 51
  religious, 39–40
paranormal cognition, 93

Parapsychological Association
  (PA), 11, 74, 75
Parapsychological Foundation, 73,
  106
parapsychology, 283, 287–89
Parisano, Emilio, 42–43
Parker, Adrian, 117
particles and waves, 211–15
Pauli, Wolfgang, 11
Pavlov's dogs, 49–50
Pearce, Hubert E. Jr., 86–89
Pearce-Pratt distance telepathy test,
  86–89
Penfield, Wilder, 251
Penrose, Sir Roger, 170–71
Persinger, Michael, 47–48, 49,
  248
personality, and presentiment,
  173–74
Personality and Individual Differences,
  40, 41
Philosophical Quarterly, The, 237
photoelectric effect, 212–13
photons, 214, 217, 227
physics:
  classical, 209–10, 211, 218
  and physical reality, 240–41
Physics World, 215
Pickering, Edward C., 64
picture-drawing tests, 89–95
  Carington, 93–95
  Sinclair, 90–92
  Warcollier, 92–93
PK effect, 154, 155
placebo effect, 61
place memories, 186
Planck, Max, 66, 212, 213
Plato, 282
Podolsky, Boris, 214
positron emission tomography
  (PET), 23
pragmatic information, model of,
  252–53
Pratt, Gaither, 86–89

precognition, 6, 81, 162; *see also* presentiment
precognitive ability, tests of, 29–31
presentiment, 161–80
    Bierman's brain test, 176–79
    common sense vs., 168
    in earthworms, 170–71
    evidence of, 179–80
    experiments on, 164–69, 170–71, 176, 203
    in forced-choice tests, 161–62
    free-running, 174–76
    in the heart, 172–73
    and personality, 173–74
    replications of studies, 169–70
    unconscious precognition tests, 163–64
Pribram, Karl, 255
Princeton Engineering Anomalies Research (PEAR), 76, 78, 95, 154–56, 158, 284, 293–94
probabilities, 64
Proceedings of the SPR, 64
pseudoscience, belief in, 38
pseudo-telepathy, 237–38, 264
psi:
    applications of, 293–95
    in autonomic nervous system, 131–33
    in central nervous system, 136–41
    coining of term, 6
    and decline effects, 268–69
    definitions of, 283–84
    in dice, 147–51
    displacement, 268, 269
    distorted in textbook descriptions, 289–90
    elusiveness of, 267–68
    as entangled minds, 3, 235, 263–66
    evidence for, 73, 275–76
    evolving concepts of, 241–44, 272–73
    feedback necessary in, 248
    field theories of, 248–49
    goal-oriented theories of, 247–48, 268
    history of, 53–59; in eighteenth century, 59–61; in nineteenth century, 61–65; in twentieth century, 66–80
    influence, 146, 289
    information transfer in, 125, 264
    limits of, 181
    missing, 268–69
    multidimensional theories of, 249–50
    as natural phenomenon, 3
    neuroscience problem in, 240
    not tightly bound to "now," 192
    opinions about, 39–41
    physics problem in, 240
    and previous performance, 173
    problems for theory development in, 240
    psychology problem for, 240
    quantum theories of, 250–60
    and quantum theory, 231–36
    in random numbers, 154–58
    signaling required in, 271
    signal-transfer theories of, 246–47
    skeptical myths debunked, 283–92
    skeptical theories of, 245–46, 277–80, 282–83
    spontaneous occurrence of, 269–70
    state of the art, 275–76
    theories of, 243–45
    unconscious, 131–40
    unsupported claims associated with, 208
psi effects, collective, 31
psi-operated switch, 78, 293–94

psi phenomena:
  broad scope of, 6
  entangled minds as, 2–3
  repeatability of, 79–80, 98
psi research, 7, 54, 68
  card-guessing experiments, 72
  decline effects in, 121–25
  on dreams, 104–9, 111, 115
  ESP testing machine in, 70
  experiment types, 81–82
  Ganzfeld studies, 115–25
  as legitimate academic study, 77
  proof-oriented and process-oriented, 104
  repeatability of, 79–80, 98–99, 102–4
  Rhine, 69–70, 83–89
  U.S. government–funded, 76–77, 78
psychic experiences:
  in communication, 237–38
  and entangled reality, 264
  profile of likely receiver of, 46
  and quantum holography, 256
psychic phenomena, 6, 232, 290
psychic surgery, 208
psychoanalysis, 60
psychokinesis (PK), 6, 82
  in dice experiments, 148–52
  as mind-matter interaction, 270–71
  PK effect, 154–55
  and random numbers, 158
Psychological Bulletin, 117, 118, 289, 290
psychology, experimental, 63
psychopathology, 51, 246
Psychophysical Research Laboratories, 76
psychophysiology, 63
psychopraxia, theory of, 248
Puthoff, Harold, 76, 95

Puységur, Armand Marie Jacques de Chastenet, Marquis de, 60, 61, 74
Pygmalion effect, 285–86

Qigong, 185
quality problem, 103–4
quanta, 212, 213
quantum computing, 15, 242
quantum cryptography, 237
quantum decoherence, 258
quantum era, birth of, 66, 242, 244
quantum holograms, 255–56
quantum ion probability clouds, 258
quantum logic, 223–24
quantum-measurement problem, 218–19
quantum-measurement process, 257
quantum mechanics, 68
quantum mind, 170
quantum physics, complementarity in, 214
quantum reality:
  as holistic, 222
  human experience in, 262
  unmediated action at a distance in, 221
quantum theories of psi, 250–60
  Bohm's implicate/explicate order, 254–56
  model of pragmatic information, 252–53
  Observational Theory, 251–52
  Stapp–von Neumann, 256–60
  Weak-Quantum Theory, 253–54
quantum theory, 6, 220–22
  applicable to complex systems, 252
  and Bell's theorem, 226, 227–31

confirmation of, 227
Copenhagen interpretation of, 222–23
and double-slit experiment, 215–20
entanglement in, 1–2, 14, 72, 76, 220, 221, 226, 235, 237–38
and EPR paper, 214, 220, 225–26
interconnectedness in, 232
many worlds interpretation of, 223
mathematics of, 214
observation in, 218, 220
physical behavior described in, 261
and psi, 231–36
role in the brain, 243
spooky action at a distance, 1, 75, 76, 221, 226, 243
wave-equation formula of, 213
Quantum Zeno Effect, 259–60

random number generators (RNGs), 154–58
and bell curve, 202
and conditioned-space hypothesis, 189
distance-dependence properties in, 191–92
and field-consciousness experiments, 182, 183–84, 185
and focused group energy, 183–84
in Global Consciousness Project, 195–201
and intention, 185
and intercorrelation value, 204, 205
and spontaneous healing, 186–91
Rao, Ramakrishna, 77

rapid eye movement (REM) sleep, 106, 107
rational empiricism, 24, 210
"rational man" mistake, 101–2
Rauscher, Elizabeth, 250
Rayleigh, Baron, 64
reaction time task, 163
reality:
affected by observation, 220, 224, 256
in "Age of Magic," 241
and Bell's inequality, 130–31
and Bell's theorem, 227–31, 238
changing assumptions about, 209, 211–12, 220–22, 238
in classical physics, 210
consciousness as fundamental ground state of, 224
Copenhagen interpretation of, 222–23, 224, 256
and decoherence, 224–25
and double-slit experiment, 215–20
and entanglement, 227, 238–39, 264
many worlds interpretation of, 223
and neorealism, 225–27
"new," 5–6, 220–22
new way of thinking about, 236–39
nonlocal connections in, 261
objective, 218
and particle-waves, 213–15
and photoelectric effect, 212–13
and pseudo-telepathy, 237–38
and psi, 231–36, 240–41
and quantum idea, 213, 218, 222
in quantum logic, 223–24
and reductionism, 221–22
role of the mind in, 219, 251

reality *(cont.)*
   tested via classical physics,
      209–10, 211, 218
   and ultraviolet catastrophe, 212
   and uncertainty, 214–15
   virtual, 224
reductionism, 221–22
Reiki, 185
religion:
   belief in, 41
   and brain activity, 47–48
   obsessions, 47
   science vs., 35
religious paranormal, 39–40
remote staring, 133–34
remote viewing, 48–49, 54, 95–96,
   256, 292–93
repressions, 31–33
research:
   autocorrelation in, 206
   bell curve in, 202
   climate in, 286
   coherence effects in, 185
   conditioned-space hypothesis
      in, 189
   confirmation bias in, 102
   controlled experiments, 35, 209
   decline effects in, 88, 121–25,
      163, 268–69
   delayed choice design in, 217
   distance-dependence properties
      in, 191–93
   effects of observation in, 220
   experimental design flaws in,
      245
   experimenter effect in, 285–87
   experimenters as separate from
      experiments, 223
   false positive in, 184
   field-consciousness experi-
      ments, 182–85, 191, 194,
      205, 207, 270
   flaw analyses in, 119
   funnel plot, 112–14

   galvanic skin response
      measured in, 83
   global mind experiments,
      194–201
   hit vs. miss approach, 129
   intercorrelation value in, 204,
      205
   interpersonal expectancy effect
      in, 285
   limitations on topics of, 7
   measurement error in, 100–102
   meta-analysis of, 102–4,
      284–85; *see also specific*
      studies
   novelty required in, 269
   participant fraud in, 115
   Pygmalion effect in, 285–86
   randomized controlled trial
      design, 24, 67
   replication of results, 79–80, 97,
      98–99, 102–4, 284
   selective reporting, 128–29,
      245–46
   statistical inference in, 64
   trial-by-trial feedback in, 127
   trim and fill, 128–29
   true negative in, 184
   weak coupling ratio in, 124
retrocognition, 247
reverberation, 248
Rhine, Joseph B., 54, 69, 92
   and ESP in animals, 74
   and ESP card tests, 72, 83–89,
      193, 288, 289
   and *Journal of Parapsychology*, 70
Rice, Tom, 39
Richet, Charles, 64, 68
Ring, Anne, 104–5
Robert-Houdin, Jean Eugene, 61
Roe, Chris, 40, 109
Römer, Hartmann, 253
Rosen, Nathan, 214
Rosenblum, Bruce, 218, 219
Rosenthal, Robert, 285–86

Rowe, William, 183–84
Rutherford, Ernest, 208

Sakaguchi, Masanobu, 78
Sako, Yoichiro, 78
Santayana, George, 79
schizotypal personality disorder,
    36–37, 41, 50
Schmidt, Helmut, 75, 248, 251
Schmidt, Stefan, 134–36
Schrödinger, Erwin, 1, 14, 213,
    223, 260
Science, 75, 136, 283
science:
    classical era, 241–42
    controlled experiments in, 35
    evidence in, 38–42, 97
    mystics vs., 233
    psychic abilities linked in, 10
    "rational man" mistake in,
        101–2
    religion vs., 35
    research in, *see* research
Science and Technology Indicators
    (NSF), 38
Scientific American, 251
scientific discovery, 296
scientific theories, power of,
    245
Scully, Marlan O., 217
séances, 61–62
second sight, 256
seeing, act of, 265
self, subliminal, 249
self-fulfilling prophecy, 285
self-observed brain, 260
self-organizing systems, 11
self-reference, 11
semantic fields theory, 249
sense of being stared at,
    125–30, 133–34
sensory awareness, 265
sensory illusions, 245
September 11 attacks:

and card-guessing test, 33–34
and Global Consciousness
    Project, 202–7
mass premonitions of, 31–33
premonitions of, 25–34
unconscious premonitions of,
    29–31
sham human circuit, 171
Shaw, Rob, 169
Sheldrake, Rupert, 127, 248–49
Sherman, Harold, 70
Shermer, Michael, 41–42, 251
Sherwood, Simon, 109
Shimony, Abner, 227
Shoup, Richard, 29, 169, 206
siddhis, 10
Sidgwick, Henry, 279–80
signal-transfer theories of psi,
    246–47
Simons, Daniel, 43
Sinclair, Mary Craig, 70, 90–91
Sinclair, Upton, 70, 90–92, 246
Sjödin, Ulf, 40
Skeptic, 41
Skeptical Inquirer, 290
skeptical theories of psi, 245–46,
    277–80, 282–83
skin-conductance responses, 164,
    168, 174–76
sleep, mind during, 277
sleep chambers, 107
sleepwalking, 60–61
Society for Psychical Research
    (SPR), 63, 64, 65, 279–80
somnambulism, 60–61
Sony Labs, 78
soul, divine presence within us,
    241
space:
    relative, 242
    transcendence of, 249–50
spirit, 241, 243
spiritual healing practices, 186–91
spiritual vibrations, 187

spontaneous healing, 186–91
spooky action at a distance, 1, 75, 76, 221, 226, 243
Spottiswoode, James, 174–75
spying, psychic, 48–49, 54
Standish, Leanna, 18–19, 137, 138
Stanford, Rex, 248
Stanford, Thomas Welton, 67
Stanford University, Psychic Fund of, 67
Stapp, Henry, 250, 256–60, 262
Stapp–von Neumann theory of psi, 256–60
STARGATE, 78
statistical inference, 64
Steinkamp, Fiona, 85, 162, 193
Stevens, Paul, 138–39
stochastic resonance phenomena, 96
Storm, Lance, 118
subliminal self, 249
Summhammer, Johann, 16, 272–73
supernatural, 55
superposition, 220
superstition, 41
Swann, Ingo, 48–49
Swedenborg, Emanuel, 59–60
synchronicity, 66

taboo topics, 7
tachyons, 247
Taft, Ryan, 186–88
Talbot, Michael, 255
Targ, Russell, 76, 95, 250
Tart, Charles, 136
technology, psi-based, 293–95
Teilhard de Chardin, Pierre, 181, 243
telekinesis, 82
telepathy, 6, 68, 81, 125, 232
    Bacon's studies of, 58–59
    brain seizures vs., 47–49
    distant mental influence research, 82–83

dream, 75
    ESP card tests, 83–89, 193
    and filtering process, 267
    Ganzfeld studies, 115–25, 289
    in hypnotic trance, 23
    Pearce-Pratt test, 86
    probabilities in, 64
    pseudo-telepathy, 237–38, 264
    and quantum theory, 72
    and signal-transfer theories, 246
    and skin-conductance responses, 164
teleportation, 208, 270
Teller, Edward, 73
temporal lobe seizures, 47–49
Tenhaeff, W. H. C., 74
Teresa, Mother, 194
Thalbourne, Michael, 248
theory, use of term, 245
Therapeutic Touch, 185
Thompson, Francis, 3–4
Thomson, Sir J. J., 65, 67
Thomson, William, 211
thought-reading, 63–64
thought transference, 63
Thouless, Robert, 6
time:
    beyond understanding, 170
    illusion of, 221
    psi not bound to "now," 192
    relative, 242
    transcendence of, 249–50, 289
Troland, Leonard, 68
truth, consensus, 279
Turing, Alan, 72–73
Turner, D. J., 280–81
twins, identical, 75
Tyrrell, G. N. M. , 70

Ullman, Montague, 75, 106–7
ultraviolet catastrophe, 212
uncertainty, 220, 252–53

uncertainty principle, 214–15
unconscious precognition tests, 163–64
unconscious psi, 131–40
U.S. government, psi research funded by, 76–77, 78

Vandegrift, Guy, 237
Vasiliev, Leonid, 74
Vassy, Zoltan, 164, 175–76
virtual reality, 224
visual evoked potentials, 140

Wackermann, Jiří, 19, 137
Walach, Harald, 253
Walker, Evan Harris, 251, 258
Walker, John, 195
Warcollier, René, 68, 92–93
Washington, George, 60
watched pot effect, 260
waves and particles, 211–15
Weak-Quantum Theory of psi, 253–54

Weiss, Ehrich (Houdini), 61
Wells, H. G., 79
Wheeler, John Archibald, 146, 217, 224
Wigner, Eugene, 224, 251
Wilber, Ken, 255
Wildey, Chester, 170–71
Wilkins, Sir Hubert, 70–72
Wiseman, Richard, 118
wishful thinking, 245
witchcraft, 41, 57, 59, 73, 74, 77
Wolfradt, Uwe, 40–41
World War I, 68
World War II, 70

Y2K, 199–201
Yamamoto, Mikio, 78
Yoga Sutras (Patanjali), 10
Young, Thomas, 212, 215
Yount, Garret, 186

Zener, Karl, 84
Zener cards, 84